MW01359759

Procedures Manual
for the
Microbiology Laboratory

Burton E. Pierce
San Diego City College

Michael J. Leboffe
San Diego City College

Morton Publishing Company
925 W. Kenyon, Unit 12
Englewood, Colorado 80110

Printed in the United States of America

10 9 8 7 6 5 4 3 2 1

ISBN: 0-89582-367-5

Recycled

Paper

Preface

A year ago, when preparing the *Photographic Atlas for the Microbiology Laboratory,* we faced an editorial decision: do we include procedures with each test or not? We decided against it, mainly because we didn't want to blur the distinction between the *Photographic Atlas* and full-fledged lab manuals. However, the most frequently encountered suggestion we received about the *Photographic Atlas* was that it would be more useful with procedures. And therein lies our incentive to produce the *Procedures Manual for the Microbiology Laboratory.*

The *Procedures Manual*, like the *Photographic Atlas*, is intended to be used as a reference rather than provide a logically sequenced curriculum. It is divided into sections that directly correspond to the nine sections in the *Photographic Atlas.* We recommend that you combine the corresponding pages of the *Procedures Manual* and the *Photographic Atlas* prior to each laboratory session, read all relevant material and examine the photos of completed tests. Make sure you understand what each result signifies and, if you have time, discuss them with your lab partners. Studying the exercises ahead of time will enable you to relax and enjoy the lab.

The basic design for each part in the *Procedures Manual* includes

- a medium recipe
- medium preparation instructions
- a materials list
- lab procedure or test/stain protocol
- precautions
- references.

We have included recipes and preparation instructions for selective, differential and general growth media used in each procedure, even though students are not likely to be responsible for its preparation. However, we believe that knowledge of ingredients is essential to understanding how the medium performs its function. It is also important to recognize that lab work for a microbiologist does *not* begin with medium inoculation.

A materials list is included for each lab. Each list is written for use by a single student, but your instructor may make adjustments and have students work in groups rather than individually. When possible we have avoided duplicating the bacteria and results which appear in the *Photographic Atlas.* In some instances we have chosen the typical positive and negative control organisms. In other cases, the organisms are less typical for a given test but, in our experience, demonstrate the appropriate biochemical activity.

Lab procedures and test or stain protocols provide step-by-step instructions on how to perform each technique. In most cases, these are fairly standardized procedures, but variations do occur (*i.e.* the Gram stain). In other cases (*i.e.,* ELISA), we have provided instructions for a kit that may be purchased. In all cases, the degree of test reproducibility is dependent upon the degree to which your test conditions meet the accepted standards. Strive for consistency and accuracy in your lab technique.

The precautions warn you of stumbling blocks you may encounter that could lead to false positive, false negative, or inconsistent results. Pay special attention to these and you should meet with better results in your lab work.

References are included for most procedures. These may guide you in further study if you want more information about a particular test or technique.

The appendix includes material necessary for you to perform the tests in the remainder of the *Procedures Manual*, but have no corresponding component in the *Photographic Atlas*. These are *not* stashed in the Appendix because they are after-thoughts. In fact, it is likely you will need at least one of the Appendix's components for most procedures you do. Appendix A provides recipes and preparation instructions for standard growth media — nutrient agar and nutrient broth. Appendix B describes various aseptic transfers. Appendix C covers microscope use.

One last specific bit of advice on how to use the *Procedures Manual*. The recurring item in Section 6 (Quantitative Techniques) is the serial dilution. In microbiology, as well as other areas of science, a working knowledge of serial dilutions and dilution factors is essential. Although the math may seem difficult at first it will become easier with time and practice. Take time to learn the procedures used in this section and pay attention to the explanations given in the *Photographic Atlas*. Ask your instructor to give you practice problems to help you become more proficient with the math used in these laboratory exercises.

It is our hope that the *Procedures Manual* will make your experience in Microbiology Laboratory more enjoyable and rewarding by providing you with background information and advice to improve your chances of success. Have a productive semester!

Burton E. Pierce
Michael J. Leboffe
August 1996

ACKNOWLEDGMENTS

We are indebted to all who assisted and advised us during the formation of this work. Thank you for your enthusiastic responses and your indispensable contributions.

James Bartley — San Diego City College
David Brady — San Diego City College
Deborah Durand — UCSD Department of Medicine
Thomas Lee — Scientific Instrument Company
William McClellan — Scientific Instrument Company
Darla Newman — San Diego Mesa College
Dr. Margaret Polley — Calbiochem-Novabiochem Corporation
Debra Reed — San Diego City College
Dr. David Singer — San Diego City College
Robert Waddell — Scientific Instrument Company
Gary Wisehart — San Diego City College

We are especially grateful to: Kay Baitz, President of Key Scientific Products, for permission to reprint the ONPG test protocol; Doreen Cantelmo of Olympus America, for the microscope photograph and permission to reproduce it; Natalie Cederquist of Hummingbird Graphics, for the excellent illustrations; Lexine Okouchi, Joanne Murphy, Jonathan Staller and Eugene Nichols from Abbott Laboratories for permission to reproduce procedural details of the ELISA test; Suzanne Graves and Sandra Olson from Abbott Laboratories for their outstanding technical advice on the ELISA protocol.

We continue to enjoy and appreciate the enthusiastic support of Doug Morton, Tom Doran and Christine Morton of Morton Publishing Company. We are also grateful to Joanne Saliger of Ash Street Typecrafters, Inc. for her design of the manual.

A special thank you goes to our wives Michele Pierce and Karen Leboffe for the love, the encouragement, and the willingness to endure the temporary inequalities inevitably produced by creative endeavors such as this.

Contents

Safety and Laboratory Guidelines

Microbiology lab can be an interesting and exciting experience, but it also has its potential for danger. As with any laboratory endeavor, improper handling of chemicals, equipment and/or microbial cultures presents a hazard to yourself and others. Listed below are some general safety rules which, if followed, will reduce the chance of injury or infection. Your instructor may supplement this list with more specific safety rules and general lab rules unique to your institution. Please follow these and any other safety guidelines required by your college.

Student Conduct

1. To reduce the risk of infection, do not smoke, eat, or drink in the laboratory room — even if lab work is not being done. Neither should you bring food or drinks into the lab, nor should you touch your mouth with your hands.
2. Wash your hands before leaving the laboratory each day. This may be done with plain soap and water, or an antiseptic soap may be provided.
3. Come to lab prepared for that day's work. Lab time is precious. Besides, figuring out what to do as you go is an undertaking designed to produce confusion and accidents.
4. Do not remove any organisms or chemicals from the laboratory.

Basic Laboratory Safety

1. Use an antiseptic (*e.g.*, Betadine) on your skin if it is exposed to a spill containing microorganisms. Your instructor will tell you which antiseptic you will be using.
2. Your instructor will tell you where the first aid kit is located.
3. Your instructor will tell you where the fire blanket and fire extinguisher are located.
4. Your instructor will tell you where the eye wash basin is located.
5. Eye protection must be worn whenever chemicals are being heated.
6. Turn off your Bunsen burner when not in use. Not only is it a fire and safety hazard, but 20-plus Bunsen burners heat up a room pretty quickly.
7. Long hair should be tied back. It is a potential source of contamination as well as a likely target to catch on fire.
8. If you are feeling ill (for whatever reason) do not work with live microbes. There are other ways you may contribute to your lab group (*i.e.*, record data, fill out culture labels, retrieve equipment, *etc.*).
9. If you are pregnant or are taking immunosuppressant drugs, please see the instructor. It may be in your best *long*-term interests to postpone taking this class.
10. Disposable latex gloves must be worn while staining microbes and handling blood products (*i.e.*, plasma, serum, antiserum, or whole blood).
11. Mouth pipetting is forbidden. Always use mechanical pipetters.
12. Uncontaminated broken glass should be disposed of in a "sharps" container.
13. All work which involves highly volatile chemicals or stains which need to be heated must be done in the fume hood.

Reducing Contamination of Self, Others, Cultures, and the Environment

1. The desk top must be wiped with a disinfectant (*e.g.*, Amphyl) before and after each lab period in which live organisms are used. Your instructor will tell you which disinfectant you will be using.
2. Culture tubes should remain upright in a tube holder. Never lay tubes on the table — even if they contain a solid medium. Condensation that forms in these tubes may leak out and contaminate the work surface, your hands, or other cultures.
3. Contain any culture spills with disinfectant-soaked towels and then report the spill to your instructor. Leave the towels on the spill for 20 minutes, then place the towels in the autoclave bag.
4. Dispose of contaminated broken glass in the container designated for autoclaving. After sterilization, the glass may be disposed of in a "sharps" container.
5. Books and papers other than what is essential for that day's lab should be kept under the desk. A cluttered lab table is an invitation for an accident, an accident that may contaminate your expensive school supplies.
6. A disinfectant-soaked towel should be placed on the work area when pipetting. This reduces the aerosols produced if a drop escapes from the pipette. It also should take care of the spill produced.
7. Used microscope slides of bacteria should be soaked in a disinfectant solution for at least 20 minutes before cleaning.

Disposal of Contaminated Materials

1. Plate cultures (if plastic Petri dishes are used) and other contaminated nonreusable items should be placed in the appropriate container to be sterilized (*e.g.*, the autoclave bag) when you are finished with them.
2. Tube cultures or any contaminated reusable items should have their labels removed and be deposited in the container designated for autoclaving.
3. Dispose of all blood product samples as well as disposable latex gloves in the container designated for autoclaving.

References

Barkley, W. Emmett and John H. Richardson. 1994. Chapter 29 in *Methods for General and Molecular Bacteriology,* edited by Philipp Gerhardt, R.G.E. Murray, Willis A. Wood, and Noel R. Kreig. American Society for Microbiology, Washington, D.C.

Collins, C.H., Patricia M. Lyne and J.M. Grange. 1995. Chapters 1 and 4 in *Collins and Lyne's Microbiological Methods, 7th Ed.* Butterworth-Heineman, Oxford.

Darlow, H. M. 1969. Chapter VI in *Methods in Microbiology, Volume 1,* edited by J. R. Norris and D. W. Ribbins. Academic Press, Ltd., London.

Fleming, Diane O. 1995. *Laboratory Safety — Principles and Practices, 2nd Ed.,* edited by Diane O. Fleming, John H. Richardson, Jerry J. Tulis and Donald Vesley. American Society for Microbiology, Washington, D.C.

Power, David A. and Peggy J. McCuen. 1988. Pages 2 and 3 in *Manual of BBL® Products and Laboratory Procedures, 6th Ed.* Becton Dickinson Microbiology Systems, Cockeysville, MD.

SECTION ONE

Bacterial Growth Patterns

Bacterial Colony Morphology

Photographic Atlas Reference

Pages 1 through 4

Procedure

1. Inoculate five plates as follows:
   a. Open one plate and expose it to the air for 30 minutes (or longer, if convenient).
   b. Use the cotton swab to sample your desk area, then streak the second agar plate as in Figure B-10a.
   c. Cough several times on the agar surface of the third plate.
   d. Touch the agar surface of the fourth agar plate lightly with your finger tips. It is best if you *haven't* washed your hands recently.
   e. Remove the lid of the fifth agar plate and vigorously scratch your head over it.
   f. Do not inoculate the sixth plate. It is a control to ensure your agar plates are sterile.
2. Label the base of each plate with your name, the date and the type of exposure it has received.
3. Incubate the six plates in an inverted position at 37°C for at least one day.
4. Figure 1-1 shows some typical colonial characteristics and their descriptive terms. Use the terms to describe colonies on your plates and those supplied to you by your instructor. You may also use the ruler to measure colony diameters. Record your results in the table below.
5. It may be helpful to use a colony counter (see Figure 1-2) to assist in viewing the plates. It has a light that shines through the agar, thus allowing you to determine optical properties of the growth. It also has a magnifying glass to allow you to see more detail.

Materials

First Lab Period

Six nutrient agar plates
A sterile cotton swab in sterile saline

Second Lab Period

A six inch plastic ruler to measure colony diameters
A colony counter
Streak plate culture of *Chromobacterium violaceum*
Streak plate culture of *Micrococcus roseus*
Streak plate culture of *Pseudomonas aeruginosa*
Streak plate culture of *Serratia marcescens*

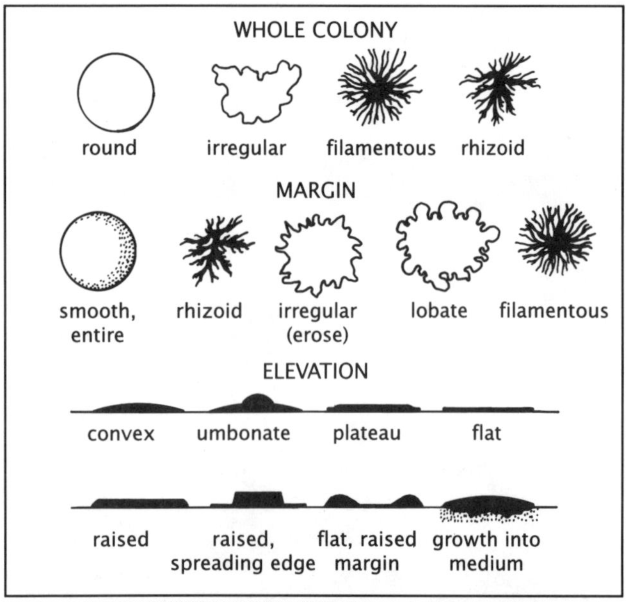

FIGURE 1-1.

A Sampling of Bacterial Colony Features *Use these terms to describe colonial morphology.*

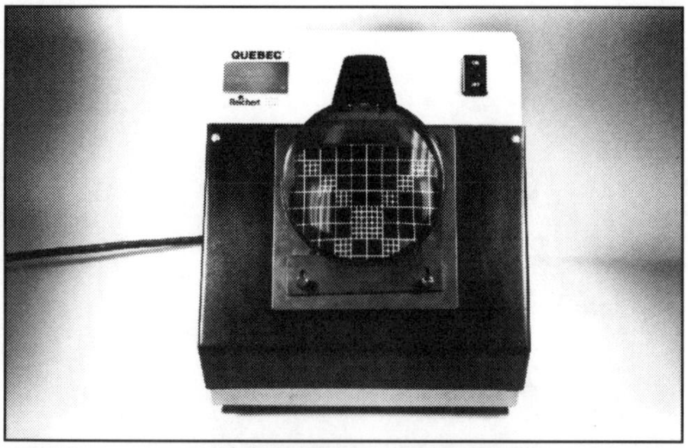

FIGURE 1-2.

The Colony Counter *The colony counter may be useful in seeing subtle differences between similar colonies. The grid in the background is composed of 1 cm squares.*

Precautions

⚠ Organisms that you find on your plates in this exercise are common laboratory contaminants, so observe them carefully.

⚠ Don't expect to see examples of all the descriptive terms in this laboratory exercise.

⚠ Save your plates for the "Growth Patterns in Broth" exercise.

References

Claus, G. William. 1989. Chapter 14 in *Understanding Microbes — A Laboratory Textbook for Microbiology.* W.H. Freeman and Company, New York, NY.

Collins, C.H., Patricia M. Lyne and J.M. Grange. 1995. Chapter 6 in *Collins and Lyne's Microbiological Methods, 7th Ed.* Butterworth-Heineman, Oxford.

Koneman, Elmer W., Stephen D. Allen, William M. Janda, Paul C. Schreckenberger and Washington C. Winn, Jr. Chapter 1 in *Color Atlas and Textbook of Diagnostic Microbiology, 4th Ed.* J.B. Lippincott Company, Philadelphia, PA.

ORGANISM/PLATE	COLONY DESCRIPTION

Growth Patterns in Broth

Photographic Atlas Reference

Page 5

Procedure

1. Using your loop and the plates from the bacterial colony morphology lab, aseptically transfer growth into five different nutrient broth tubes.
2. Label each tube with your name, the date, the medium and a description of the colony (*e.g.*, small, white circular colony).
3. Incubate the tubes at 37°C for at least one day.
4. After incubation, examine the tubes and describe differences in growth. Record the results in the table below. Page 5 in the *Atlas* lists some of the descriptive terms you should use.

Materials

Five nutrient broth tubes
Your plates from the colony morphology lab

Precautions

⚠ Check with your classmates to see if they got growth patterns consistent with those of your organisms.

⚠ You may not see examples of all growth patterns from your samples. Check to see if your classmates got any different ones.

⚠ You may see more than one pattern in a single tube, as with *Chromobacterium violaceum* in the *Atlas* which illustrates a ring, sediment and fine turbidity.

Reference

Claus, G. William. 1989. Chapter 17 in *Understanding Microbes — A Laboratory Textbook for Microbiology.* W.H. Freeman and Company, New York, NY.

ORGANISM	DESCRIPTION OF GROWTH IN BROTH

Aerotolerance — Agar Deep Stabs

Photographic Atlas Reference

Page 6

Test Protocol

1. Use your inoculating needle to stab the agar of three tubes with different organisms. Use a heavy inoculum.

Materials

Four nutrient agar deep tubes
18 to 24 hour pure cultures of:
Clostridium spp. in thioglycolate broth
Moraxella spp. in nutrient broth
Providencia spp. in nutrient broth

2. Stab the fourth tube with your sterile needle to act as a control.
3. Label each tube with your name, the date, and the medium.
4. Incubate the tubes at 37°C for at least one day.
5. After incubation, examine the tubes and try to determine the aerotolerance category of each. Record the results in the table below.

Precautions

⚠ Use your control tube to help you discriminate between growth along the stab line and the stab line itself.

⚠ Try to enter and exit along the same stab line.

ORGANISM	DESCRIPTION OF GROWTH	AEROTOLERANCE CATEGORY
Clostridium spp.		
Moraxella spp.		
Providencia spp.		

Anaerobic Culture Methods — Thioglycolate Broth

Photographic Atlas Reference

Page 7

Recipe

Thioglycolate Broth

Yeast Extract	5.0 g
Casitone	15.0 g
Dextrose (glucose)	5.5 g
Sodium chloride	2.5 g
Sodium thioglycolate	0.5 g
L-Cystine	0.5 g
Agar	0.75 g
Resazurin	0.001 g
Distilled or deionized water	1.0 L

final pH = 7.1 ± 0.2 at 25°C

Medium Preparation

1. Suspend the ingredients in one liter of distilled or deionized water. Boil to completely dissolve them.
2. Dispense 10.0 mL into sterile test tubes.
3. Autoclave for 15 minutes at 15 lbs. pressure (121°C) to sterilize.

Materials

Four thioglycolate tubes
18 to 24 hour pure cultures of:
- *Clostridium sporogenes*
- *Pseudomonas aeruginosa*
- *Staphylococcus aureus*

Test Protocol

1. Inoculate three thioglycolate tubes with the test organisms. Use the fourth tube as an uninoculated control.
2. Label each tube with your name, the date, medium and organism.
3. Incubate the tubes at 37°C for at least one day.
4. After incubation, examine the tubes and determine the oxygen tolerance category of each. Record the results in the table below.

Precautions

⚠ Thioglycolate tubes should have a pink band at the surface (the aerobic zone). If none is visible, the medium is of suspect quality.

⚠ If after storage, more than 30% of the medium is pink, the tubes should be boiled with caps loosened to remove the excess oxygen. Tighten the caps and cool to room temperature before use.

References

DIFCO Laboratories. 1984. Page 951 in *DIFCO Manual, 10th Ed.* DIFCO Laboratories, Detroit, MI.

Power, David A. and Peggy J. McCuen. 1988. Page 261 in *Manual of BBL® Products and Laboratory Procedures, 6th Ed.* Becton Dickinson Microbiology Systems, Cockeysville, MD.

ORGANISM	DESCRIPTION OF GROWTH IN BROTH	OXYGEN TOLERANCE CATEGORY
Clostridium sporogenes		
Pseudomonas aeruginosa		
Staphylococcus aureus		

Anaerobic Culture Methods — Anaerobic Jar

Photographic Atlas Reference

Page 7

Test Protocol

1. Divide each nutrient agar plate into three sectors with your marking pen.
2. Use your loop to spot inoculate the three organisms into different sectors of each plate.
3. Label each plate with your name, the organisms' names in each sector, the date, medium and incubation conditions.
4. Place *one* plate in the anaerobic jar in an inverted position.
5. When the jar is filled, discharge the packet as follows (or follow the instructions on your packet).
   a. Stick the methylene blue strip on the wall of the jar.
   b. Open the packet and dispense 10 mL of distilled water.
   c. Place the open packet with the writing facing inward into the jar.
   d. Immediately close the jar.
6. Incubate the anaerobic jar at 37°C for at least one day. Incubate the other plate aerobically at the same temperature for the same length of time.
7. After incubation, examine and compare the plates. Record the results in the table below.

Materials

Two nutrient agar plates
One anaerobic jar with gas generator packet
18 to 24 hour pure cultures of:
Clostridium sporogenes
Pseudomonas aeruginosa
Staphylococcus aureus

Precautions

⚠ Make certain the gas generator packet for the anaerobic jar has not expired.

⚠ Once water is added to the gas generator packet, close the lid immediately.

⚠ If no condensation is seen inside the jar within 30 minutes, check the seal on the jar lid and repeat the process.

⚠ After incubation, check the methylene blue strip before opening the jar. It should be white. If it is not, conditions may not have been anaerobic.

References

Allen, Stephen D., Jean A. Siders, and Linda M. Marler. 1985. Chapter 37 in *Manual of Clinical Microbiology, 4th Ed.* Edited by Edwin H. Lennette, Albert Balows, William J. Hausler, Jr., and H. Jean Shadomy. American Society for Microbiology, Washington, D.C.

Power, David A. and Peggy J. McCuen. 1988. Page 311 in *Manual of BBL® Products and Laboratory Procedures, 6th Ed.* Becton Dickinson Microbiology Systems, Cockeysville, MD.

ORGANISM	AEROBIC GROWTH (+ or –)	ANAEROBIC GROWTH (+ or –)	OXYGEN TOLERANCE CATEGORY
Clostridium sporogenes			
Pseudomonas aeruginosa			
Staphylococcus aureus			

SECTION TWO

Isolation Techniques and Selective Media

Streak Plate Method of Isolation

Photographic Atlas Reference

Page 9

Test Protocol

1. The quadrant method of isolation is performed as illustrated in Figures 2-1a through 2-1d. In this exercise, you will be streaking nutrient agar. However, in a clinical laboratory, the streak technique is also used in conjunction with selective media such as the ones that comprise the remainder of this section.
2. Before you attempt to streak on a real agar plate, you may practice the wrist action using a pencil on the "plate" shown.
3. Streak the mixture of *Chromobacterium violaceum* and *Escherichia coli* on one of your nutrient agar plates. Label the plate with your name, the date, and the organisms.
4. Streak the mixture *Enterobacter aerogenes* and *Escherichia coli* on the other nutrient agar plate. Label the plate with your name, the date, and the organisms.

Materials

Two nutrient agar plates per student

Broth culture of *Chromobacterium violaceum* and *Escherichia coli* mixed together in equal portions immediately before use

Broth culture of *Enterobacter aerogenes* and *Escherichia coli* mixed together in equal portions immediately before use

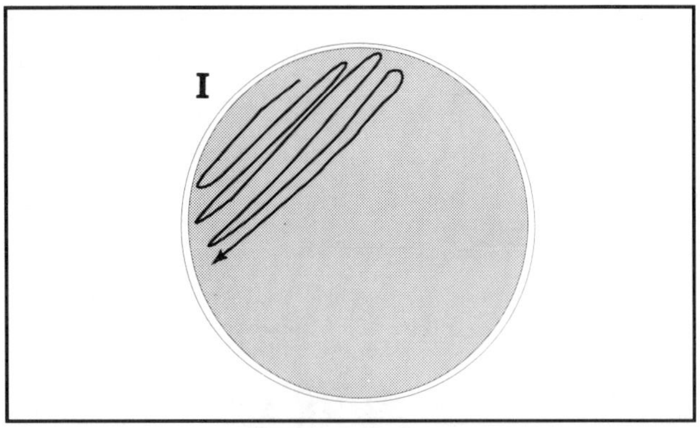

FIGURE 2-1a.

Beginning the Streak Pattern *Streak the mixed culture back and forth in one quadrant of the agar plate. Use the lid as a shield. Flame the loop, then proceed.*

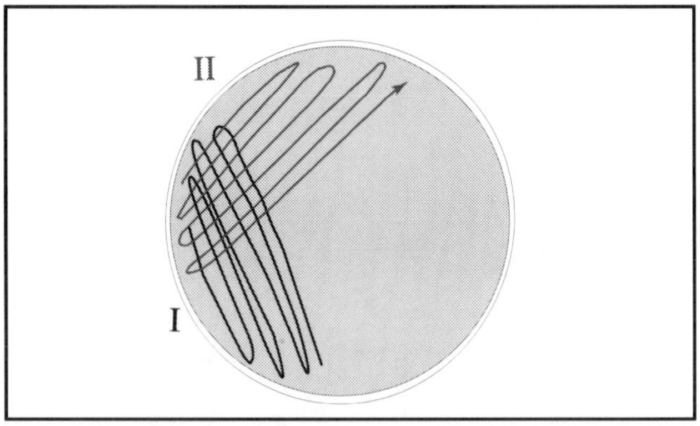

FIGURE 2-1b.

Streak Again *Rotate the plate 90˚ and touch the agar in an uninoculated region to cool the loop. Streak again using the same wrist motion. Flame the loop.*

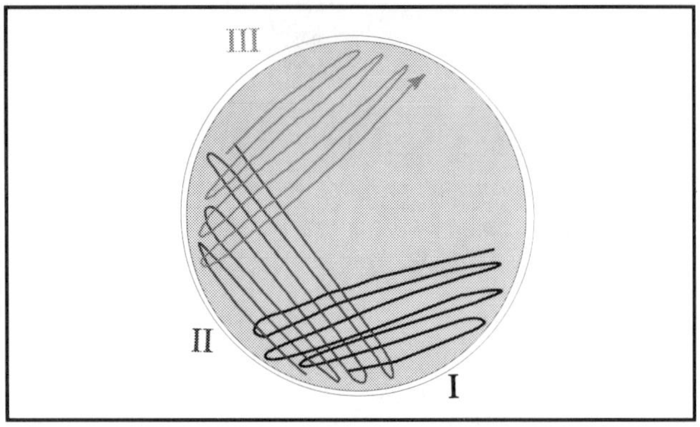

FIGURE 2-1c.

Streak Yet Again *Rotate the plate 90˚ and streak again using the same wrist motion. Be sure to cool the loop prior to streaking. Flame again.*

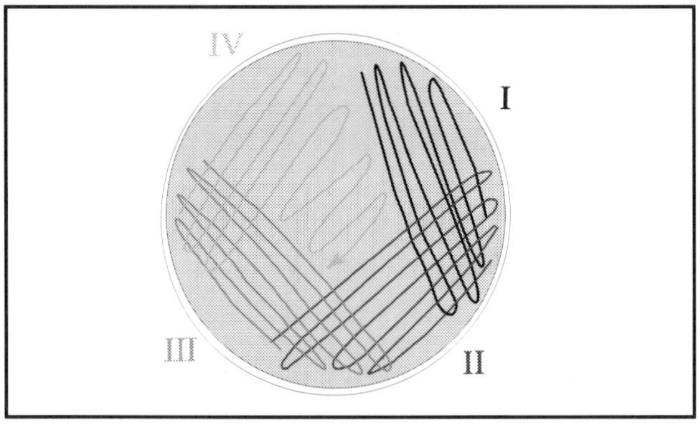

FIGURE 2-1d.

Streak Into the Center *After cooling the loop, streak one last time into the center of the plate. Flame the loop and incubate the plate in an inverted position for the assigned time at the appropriate temperature.*

5. Tape the two plates together and incubate them in an inverted position for at least 24 hours at 37°C.
6. After incubation, observe the plates for isolation. Examine the streak pattern to determine in which streak you achieved isolation. If no isolation was obtained, try to determine if technique was responsible.

Precautions

⚠ Avoid cutting into the agar surface with the loop as you streak.

⚠ Flame the loop between streaks and especially when you are finished.

⚠ Be careful not to use a hot loop for streaking as you will create aerosols and may kill the organism you are trying to isolate.

⚠ Be sure to rotate the plate 90° between each streak and pass through the preceding streak pattern, otherwise you'll have no organisms to streak!

⚠ Be aware of the plate's rotation direction during the procedure. If you rotate clockwise the first time and counterclockwise thereafter, you will not likely have a successful streak plate.

⚠ Since isolated colonies is the goal of this procedure, be sure to incubate the plate in an inverted position. If you don't, condensation dropping from the lid will wash across the agar surface and the growth will not be in distinct colonies.

References

Baron, Ellen Jo, Lance R. Peterson and Sydney M. Finegold. 1994. Chapter 9 in *Bailey and Scott's Diagnostic Microbiology, 9th Ed.* Mosby-Yearbook, St. Louis, MO.

Collins, C. H. and Patricia M. Lyne. 1995. Chapter 6 in *Collins and Lyne's Microbiological Methods, 7th Ed.* Butterworth-Heineman.

Koneman, Elmer W., Stephen D. Allen, William M. Janda, Paul C. Schreckenberger and Washington C. Winn, Jr. Chapter 1 in *Color Atlas and Textbook of Diagnostic Microbiology, 4th Ed.* J.B. Lippincott Company, Philadelphia, PA.

Power, David A. and Peggy J. McCuen. 1988. Pages 2 and 3 in *Manual of BBL® Products and Laboratory Procedures, 6th Ed.* Becton Dickinson Microbiology Systems, Cockeysville, MD.

Desoxycholate (DOC) Agar

Photographic Atlas Reference

Page 10

Recipe

Modified Leifson Desoxycholate Agar

Peptone	10.0 g
Lactose	10.0 g
Sodium desoxycholate	1.0 g
Sodium chloride	5.0 g
Dipotassium phosphate	2.0 g
Ferric citrate	1.0 g
Sodium citrate	1.0 g
Agar	16.0 g
Neutral red	0.033 g
Distilled or deionized water	1.0 L

final pH = 7.3 ± 0.2 at 25°C

Medium Preparation

1. Heat and stir the ingredients in one liter of distilled or deionized water. Boil for 1 minute to make certain they are completely dissolved.
2. When cooled to 50°C, pour into sterile plates.
3. Allow to cool to room temperature.

Materials

One DOC plate

18 to 24 hour pure cultures of:

Escherichia coli
Klebsiella pneumoniae
Staphylococcus aureus

Test Protocol

1. Use a marking pen to divide the plate into four sectors. Marks should be made on the plate's base.
2. Spot inoculate in different sectors and leave the fourth sector uninoculated as a control. (Spot inoculations are for demonstration only. This medium is intended to be streaked for isolation.)
3. Label the plate with the organisms' names, your name, and the date.
4. Invert the plate and incubate at 35°C for 24 to 48 hours.
5. Examine the plate and record your results below.

Precaution

⚠ The medium is heat sensitive. It should not be autoclaved or overheated when prepared, nor can it be remelted and repoured.

References

DIFCO Laboratories. 1984. Page 274 in *DIFCO Manual, 10th Ed.* DIFCO Laboratories, Detroit, MI.

Power, David A. and Peggy J. McCuen. 1988. Page 144 in *Manual of BBL® Products and Laboratory Procedures, 6th Ed.* Becton Dickinson Microbiology Systems, Cockeysville, MD.

ORGANISM	RESULT
Escherichia coli	
Klebsiella pneumoniae	
Staphylococcus aureus	

Eosin Methylene Blue (EMB) Agar

Photographic Atlas Reference

Page 11

Recipe

Eosin Methylene Blue Agar

Peptone	10.0 g
Lactose	10.0 g*
Dipotassium phosphate	2.0 g
Agar	15.0 g
Eosin Y	0.4 g
Methylene blue	0.065 g
Distilled or deionized water	1.0 L

final pH = 7.1 ± 0.2 at 25°C

*An alternate recipe replaces the 10.0 g of lactose with 5.0 g of lactose and 5.0 g of sucrose.

Medium Preparation

1. Mix and heat the ingredients in one liter of distilled or deionized water until they are completely dissolved.
2. Autoclave for 15 minutes at 15 lbs. pressure (121°C).
3. When cooled to 50°C, pour into sterile plates.
4. Allow to cool to room temperature.

Materials

One EMB plate

18 to 24 hour pure cultures of:

Escherichia coli
Klebsiella pneumoniae
Staphylococcus aureus

Test Protocol

1. Use a marking pen to divide the plate into three sectors. Marks should be made on the plate's base.
2. Spot inoculate in different sectors and leave the fourth sector uninoculated as a control. (Spot inoculations are for demonstration only. This medium is intended to be streaked for isolation.)
3. Label the plate with the organisms' names, your name, and the date.
4. Invert the plate and incubate at 35°C for 24 to 48 hours.
5. Examine the plate and record your results below.

Precaution

⚠ Negative results after 24 hours should be incubated for another 24 hours.

References

DIFCO Laboratories. 1984. Page 324 in *DIFCO Manual, 10th Ed.* DIFCO Laboratories, Detroit, MI.

Power, David A. and Peggy J. McCuen. 1988. Page 153 in *Manual of BBL® Products and Laboratory Procedures, 6th Ed.* Becton Dickinson Microbiology Systems, Cockeysville, MD.

ORGANISM	RESULT
Escherichia coli	
Klebsiella pneumoniae	
Staphylococcus aureus	

MacConkey Agar

Photographic Atlas Reference

Page 12

Recipe

MacConkey Agar

Pancreatic digest of gelatin	17.0 g
Pancreatic digest of casein	1.5 g
Peptic digest of animal tissue	1.5 g
Lactose	10.0 g
Bile salts	1.5 g
Sodium chloride	5.0 g
Neutral red	0.03 g
Crystal violet	0.001g
Agar	13.5 g
Distilled or deionized water	1.0 L

final pH = 7.1 ± 0.2 at 25°C

Medium Preparation

1. Mix and heat the ingredients in one liter of distilled or deionized water until they are dissolved. Boil for 1 minute to make certain they are completely dissolved.
2. Autoclave for 15 minutes at 15 lbs. pressure (121°C).
3. When cooled to 50°C, pour into sterile plates.
4. Allow to cool to room temperature.

Materials

One MacConkey Agar plate

18 to 24 hour pure cultures of:

Enterobacter aerogenes
Proteus vulgaris
Staphylococcus aureus

Test Protocol

1. Use a marking pen to divide the plate into four sectors. Marks should be made on the plate's base.
2. Spot inoculate in different sectors and leave the fourth sector uninoculated as a control. (Spot inoculations are for demonstration only. This medium is intended to be streaked for isolation.)
3. Label the plate with the organisms' names, your name, and the date.
4. Invert the plate and incubate at 35°C for 24 to 48 hours.
5. Examine the plate and record your results below.

Precautions

⚠ The agar surface should be dry prior to inoculation.

⚠ Negative results after 24 hours should be incubated for another 24 hours.

References

DIFCO Laboratories. 1984. Page 546 in *DIFCO Manual, 10th Ed.* DIFCO Laboratories, Detroit, MI.

Power, David A. and Peggy J. McCuen. 1988. Page 189 in *Manual of BBL® Products and Laboratory Procedures, 6th Ed.* Becton Dickinson Microbiology Systems, Cockeysville, MD.

ORGANISM	RESULT
Enterobacter aerogenes	
Proteus vulgaris	
Staphylococcus aureus	

Mannitol Salt Agar (MSA)

Photographic Atlas Reference

Page 13

Recipe

Mannitol Salt Agar

Beef extract	1.0 g
Peptone	10.0 g
Sodium chloride	75.0 g
D-Mannitol	10.0 g
Phenol red	0.025 g
Agar	15.0 g
Distilled or deionized water	1.0 L

final pH = 7.4 ± 0.2 at 25°C

Medium Preparation

1. Suspend the ingredients in one liter of distilled or deionized water and mix. Boil one minute to completely dissolve ingredients.
2. Autoclave for 15 minutes at 15 lbs. pressure (121°C).
3. When cooled to 50°C, pour into sterile plates.
4. Allow plates to cool to room temperature.

Materials

One Mannitol Salt Agar plate

18 to 24 hour pure cultures of:

Staphylococcus aureus
Staphylococcus epidermidis

Test Protocol

1. Use a marking pen to divide the plate into four sectors. Marks should be made on the plate's base.
2. Spot inoculate in different sectors and leave the third sector uninoculated as a control. (Spot inoculations are for demonstration only. This medium is intended to be streaked for isolation.)
3. Label the plate with the organisms' names, your name, and the date.
4. Invert the plate and incubate at 35°C for 24 to 48 hours.
5. Examine the plate and record your results below.

Precaution

⚠ Use caution examining plates if you streaked from a patient's sample since this medium selects for pathogenic staphylococci.

References

DIFCO Laboratories. 1984. Page 558 in *DIFCO Manual, 10th Ed.* DIFCO Laboratories, Detroit, MI.

Power, David A. and Peggy J. McCuen. 1988. Page 193 in *Manual of BBL® Products and Laboratory Procedures, 6th Ed.* Becton Dickinson Microbiology Systems, Cockeysville, MD.

ORGANISM	RESULT
Staphylococcus aureus	
Staphylococcus epidermidis	

Phenylethyl Alcohol (PEA) Agar

Photographic Atlas Reference

Page 14

Recipe

Phenylethyl Alcohol Agar

Tryptose	10.0 g
Beef extract	3.0 g
Sodium chloride	5.0 g
Phenylethyl alcohol	2.5 g
Agar	15.0 g
Distilled or deionized water	1.0 L

final pH = 7.3 ± 0.2 at 25°C

Medium Preparation

1. Suspend the ingredients in one liter of distilled or deionized water. Boil one minute to completely dissolve ingredients.
2. Autoclave for 15 minutes at 15 lbs. pressure (121°C).
3. When cooled to 50°C, pour into sterile plates.
4. Allow plates to cool to room temperature.

Materials

One PEA plate

18 to 24 hour pure cultures of:

Enterococcus faecalis
Proteus mirabilis
Staphylococcus epidermidis

Test Protocol

1. Use a marking pen to divide the plate into four sectors. Marks should be made on the plate's base.
2. Spot inoculate in different sectors and leave the fourth sector uninoculated as a control. (Spot inoculations are for demonstration only. This medium is intended to be streaked for isolation.)
3. Label the plate with the organisms' names, your name, and the date.
4. Invert the plate and incubate at 35°C for 24 to 48 hours.
5. Examine the plate and record your results below.

Precaution

⚠ PEA agar is designed to isolate gram-positive microbes, especially gram-positive cocci, by inhibiting gram-negatives. Some gram-negatives, however, are only partially inhibited and may grow slowly on this medium.

References

DIFCO Laboratories. 1984. Page 667 in *DIFCO Manual, 10th Ed.* DIFCO Laboratories, Detroit, MI.

Power, David A. and Peggy J. McCuen. 1988. Page 223 in *Manual of BBL® Products and Laboratory Procedures, 6th Ed.* Becton Dickinson Microbiology Systems, Cockeysville, MD.

ORGANISM	RESULT
Enterococcus faecalis	
Proteus mirabilis	
Staphylococcus epidermidis	

Xylose Lysine Desoxycholate (XLD) Agar

Photographic Atlas Reference

Page 15

Recipe

Xylose Lysine Desoxycholate Agar

Xylose	3.5 g
L-Lysine	5.0 g
Lactose	7.5 g
Sucrose	7.5 g
Sodium chloride	5.0 g
Yeast extract	3.0 g
Phenol red	0.08 g
Sodium desoxycholate	2.5 g
Sodium thiosulfate	6.8 g
Ferric ammonium citrate	0.8 g
Agar	13.5 g
Distilled or deionized water	1.0 L

final pH = 7.5 ± 0.2 at 25°C

Preparation

1. Suspend the ingredients in one liter of distilled or deionized water and mix. Heat only until the medium boils.
2. Cool in a water bath at 50°C.
3. When cooled, pour into sterile plates.
4. Allow the plates to cool to room temperature.

Materials

One XLD plate

18 to 24 hour pure cultures of:

Enterobacter aerogenes
Providencia stuartii
Salmonella typhimurium

Test Protocol

1. Use a marking pen to divide the plate into four sectors. Marks should be made on the plate's base.
2. Spot inoculate in different sectors and leave the fourth sector uninoculated as a control. (Spot inoculations are for demonstration only. This medium is intended to be streaked for isolation.)
3. Label the plate with the organisms' names, your name, and the date.
4. Invert the plate and incubate at 35°C for 18 to 24 hours.
5. Examine the plate and record your results below.

Precautions

⚠ Avoid overheating the medium while preparing it and do not autoclave it. Heat causes a precipitate to form.

⚠ Incubation periods longer than 24 hours may lead to false identification of *Shigella* species, which are distinguished from other enterics by their inability to ferment xylose—and so produce a red color on XLD. Some microbes ferment the carbohydrates in XLD to acid and lower the pH, then decarboxylate the lysine with a subsequent raising of the pH which produces a red color.

References

DIFCO Laboratories. 1984. Page 1128 in *DIFCO Manual, 10th Ed.* DIFCO Laboratories, Detroit, MI.

Power, David A. and Peggy J. McCuen. 1988. Page 288 in *Manual of BBL® Products and Laboratory Procedures, 6th Ed.* Becton Dickinson Microbiology Systems, Cockeysville, MD.

ORGANISM	RESULT
Enterobacter aerogenes	
Providencia stuartii	
Salmonella typhimurium	

Bacterial Cellular Morphology and Simple Stains

Negative Stain

Photographic Atlas Reference

Page 17

Staining Protocol

1. Figures 3-1a through 3-1f illustrate the procedure for preparing a negative stain.

Materials

Nigrosin stain
Clean glass microscope slides
Disposable latex gloves
18 to 24 hour nutrient agar slant pure cultures of:
Bacillus megaterium
Micrococcus luteus

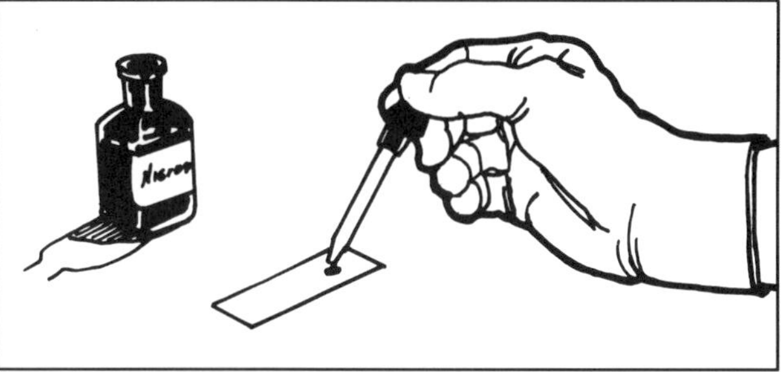

FIGURE 3-1a.

Begin with the Stain *Place a small drop of nigrosin stain at one end of a clean glass slide. Avoid excess nigrosin on the slide. It is advisable to wear latex gloves to protect your hands.*

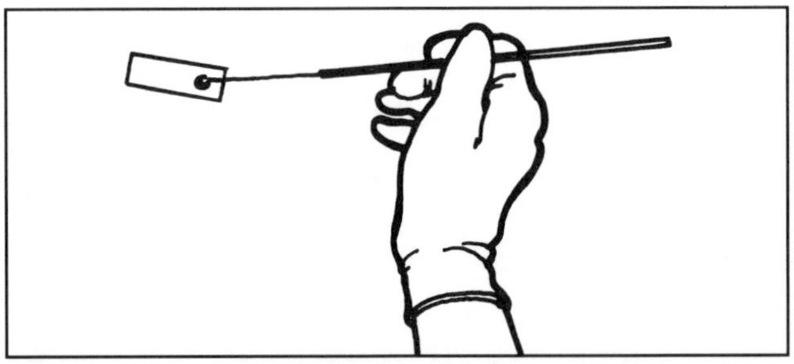

FIGURE 3-1b.

Add the Organism *Use a loop to aseptically transfer cells to the slide. Gently mix the organisms into the nigrosin. Avoid over-inoculating the slide or spattering the contaminated nigrosin drop as you mix. Flame the loop before proceeding.*

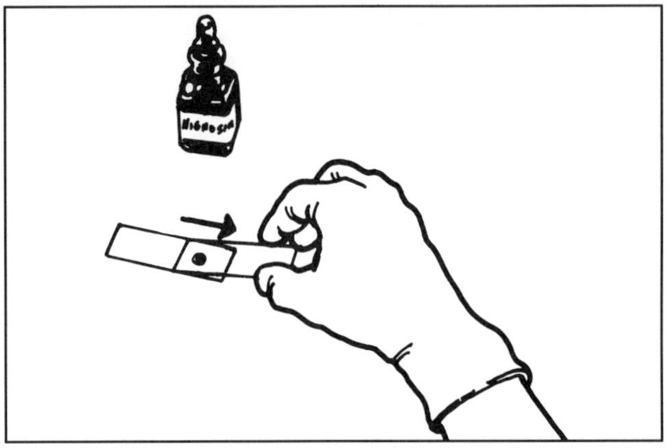

FIGURE 3-1c.

Use a Second Slide as a Spreader *Place a clean second slide on the surface of the first slide and carefully back it up into the drop of nigrosin.*

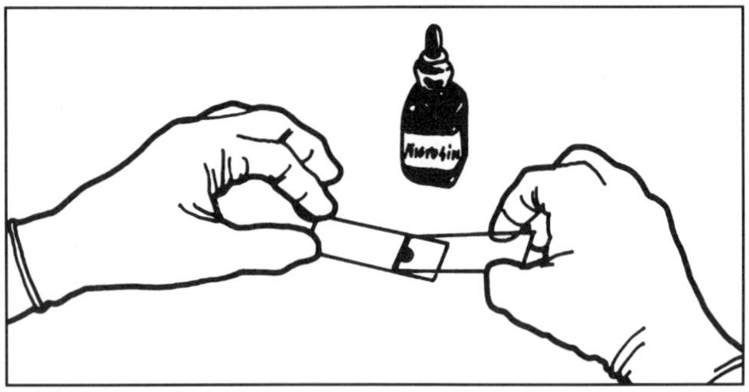

FIGURE 3-1d.

Not Too Far! *As soon as the nigrosin flows across the width of the spreader slide, stop.*

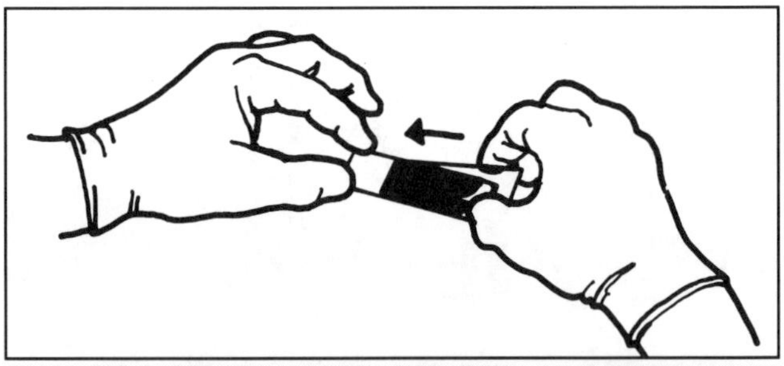

FIGURE 3-1e.

Spread a Nigrosin Film Across the Slide *Make a nigrosin smear by pushing the spreader slide across the specimen slide's surface. Allow the film to air dry. Dispose of the spreader slide appropriately since it is contaminated.*

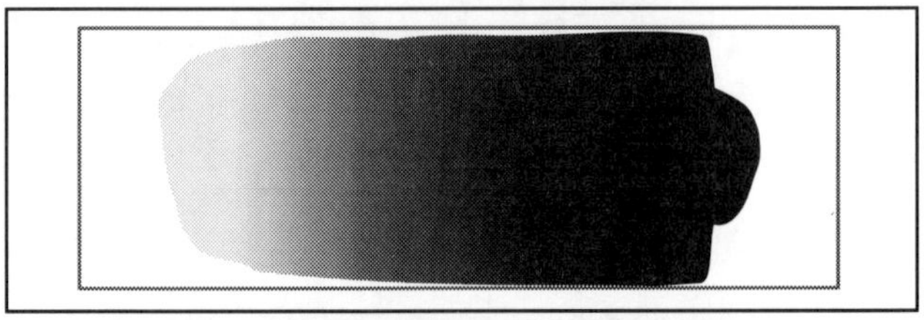

FIGURE 3-1f.

The Finished Product *After air drying, the slide should look like this. Use the oil lens to observe a region where the nigrosin is relatively thin.*

2. Observe using the oil immersion lens. Record your results in the table below.

Precautions

⚠ Spread the nigrosin as soon as it flows across the width of the spreader slide. Too much nigrosin in the smear makes microscopic observation difficult.

⚠ Dispose of the spreader slide in a jar of disinfectant immediately after use.

⚠ Dispose of the specimen slide in a jar of disinfectant after use.

References

Claus, G. William. 1989. Chapter 11 in *Understanding Microbes — A Laboratory Textbook for Microbiology*. W.H. Freeman and Company, New York, NY.

Murray, R.G.E., Raymond N. Doetsch and C.F. Robinow. 1994. Page 27 in *Methods for General and Molecular Bacteriology*, edited by Philipp Gerhardt, R.G.E. Murray, Willis A. Wood, and Noel R. Krieg. American Society for Microbiology, Washington, D.C.

ORGANISM	CELLULAR MORPHOLOGY AND ARRANGEMENT	CELL DIMENSIONS
Bacillus megaterium		
Micrococcus luteus		

Preparing a Bacterial Smear

Photographic Atlas Reference

None

Protocol

1. A bacterial smear is made prior to most staining procedures. Figures 3-2a through 3-2d illustrate the procedure for preparing a bacterial smear.
2. Follow with the appropriate staining procedure.

Precautions

⚠ Do not use too much water in preparing the slide. This will prolong the air drying step.

⚠ Do not over-inoculate the smear. When dry, it should be barely visible as a film on the glass.

⚠ Although air drying is the most time-consuming step in the procedure, resist the temptation to wave the slide, blow on it, or heat it to speed up the drying process, as contaminating aerosols may result.

⚠ Do not overheat the slide while heat-fixing it. You should be able to handle a properly heat-fixed slide without burning yourself.

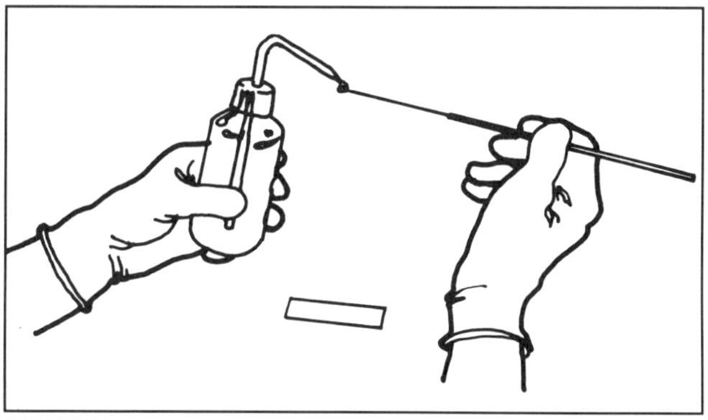

FIGURE 3-2a.

Begin With Water *Capture a drop of water with your inoculating loop. If your specimen is growing in broth, you may omit the water drop and continue with step 3-2c.*

Materials

Bacterial culture (as assigned)
Clean glass microscope slides
Inoculating loop or needle

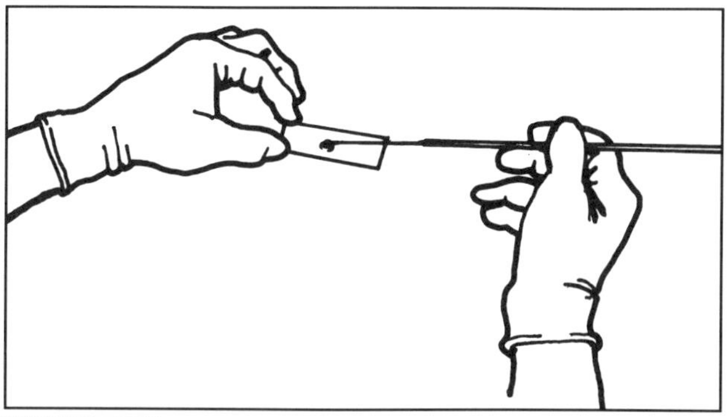

FIGURE 3-2b.

Place the Water on the Slide *Transfer the water drop to the center of a clean slide. Avoid using too much water.*

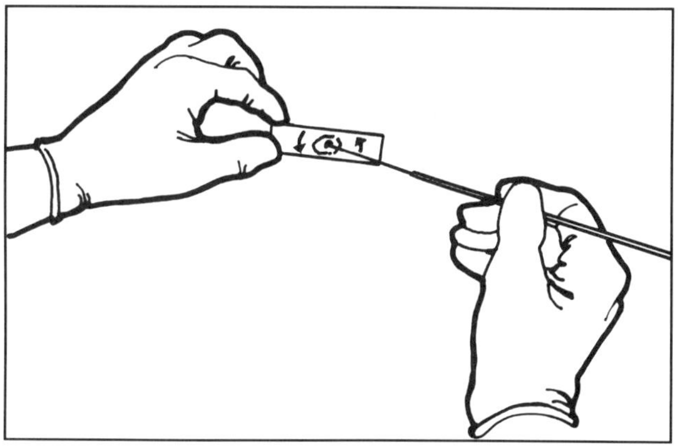

FIGURE 3-2c.

Transfer the Organisms to the Slide *Use your loop to aseptically transfer the cells to the water drop. Avoid excessive inoculation as thick smears are more difficult to stain consistently and may peel off the slide. Then, without "springing" your loop, gently mix (emulsify) the cells in the drop. As you do so, spread the drop out over the slide's surface so it will air dry more quickly. The slide must be completely dry before continuing.*

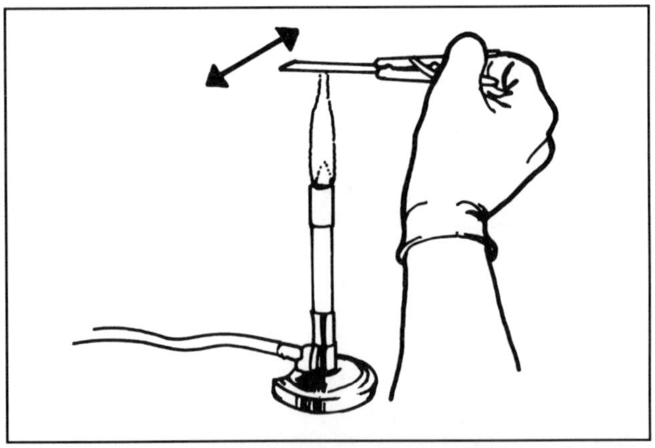

FIGURE 3-2d.

Heat-Fix the Slide *Once the drop has air dried, use a slide holder and pass the slide through the upper part of a flame two or three times to heat-fix the smear. Heat-fixing the dried smear makes the cells adhere to the slide, kills them, and makes them more easily stained as protein becomes coagulated. Do not overheat the slide as aerosols may be produced.*

References

Chapin, Kimberle. 1995. Chapter 4 in *Manual of Clinical Microbiology, 6th Ed.*, edited by Patrick R. Murray, Ellen Jo Baron, Michael A. Pfaller, Fred C. Tenover, and Robert H. Yolken. American Society for Microbiology, Washington, D.C.

Claus, G. William. 1989. Chapter 5 in *Understanding Microbes — A Laboratory Textbook for Microbiology*. W.H. Freeman and Company, New York, NY.

DIFCO Laboratories. 1984. *DIFCO Manual, 10th Ed.* DIFCO Laboratories, Detroit, MI.

Murray, R.G.E., Raymond N. Doetsch and C.F. Robinow. 1994. Page 27 in *Methods for General and Molecular Bacteriology*, edited by Philipp Gerhardt, R.G.E. Murray, Willis A. Wood, and Noel R. Krieg. American Society for Microbiology, Washington, D.C.

Norris, J. R. and Helen Swain. 1971. Chapter II in *Methods in Microbiology, Volume 5A*, edited by J. R. Norris and D. W. Ribbons. Academic Press, Ltd., London.

Power, David A. and Peggy J. McCuen. 1988. Page 4 in *Manual of BBL® Products and Laboratory Procedures, 6th Ed.* Becton Dickinson Microbiology Systems, Cockeysville, MD.

Simple Stain

Photographic Atlas Reference

Page 18

Staining Protocols

1. Prepare and heat-fix a smear of each organism.
2. Follow the basic staining procedure illustrated in Figures 3-3a through 3-3c. Prepare two slides with each stain using the following times:

    Crystal violet: stain for 30 to 60 seconds
    Safranin: stain for up to 1 minute
    Methylene blue: stain for 30 to 60 seconds

    Record the duration of staining in the table below so you can adjust staining time if the specimen is overstained or understained.
3. Observe using the oil immersion lens. Record your observations of cell morphology, arrangement and size in the table provided.

Materials

Clean glass microscope slides
Methylene blue stain
Safranin stain
Crystal violet stain
Disposable latex gloves
18 to 24 hour nutrient agar slant pure cultures of:

Bacillus cereus
Micrococcus luteus
Moraxella catarrhalis
Rhodospirillum rubrum
Staphylococcus epidermidis
Vibrio harveyi

FIGURE 3-3a.

Flood the Smear with Stain *Place your slide with its smear on the staining rack. Flood the smear with stain for the correct length of time. Hold the slide with a slide holder to minimize staining of your hands. Wearing latex gloves is also a good idea.*

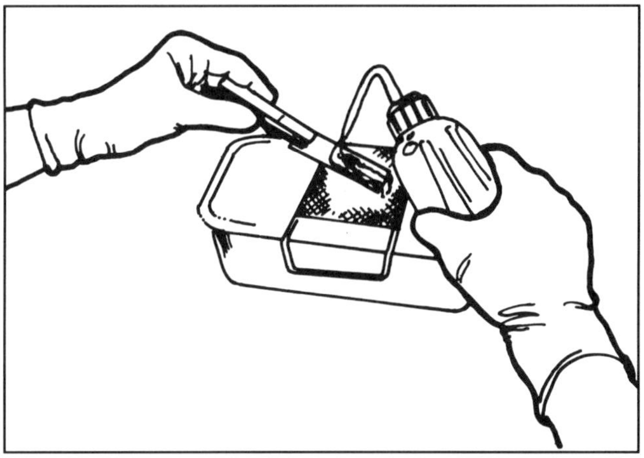

FIGURE 3-3b.

Rinse with Water *Tilt the slide to an angle of 45°. Direct a stream of water towards the top of the slide and allow the water to run down across the smear's surface. Continue washing until the runoff is clear.*

ORGANISM	STAIN AND DURATION	CELLULAR MORPHOLOGY AND ARRANGEMENT	CELL DIMENSIONS
Bacillus cereus			
Micrococcus luteus			
Moraxella catarrhalis			
Rhodospirillum rubrum			
Staphylococcus epidermidis			
Vibrio harveyi			

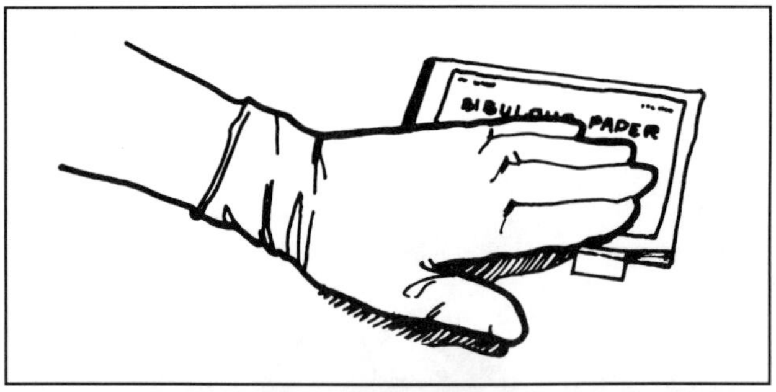

FIGURE 3-3c.

Blot Dry *Blot (do not wipe) the slide dry in a tablet of bibulous paper. Then observe the specimen using the oil immersion lens.*

Precautions

⚠ Preparation of smears with a consistent cell density makes producing slides with consistent staining results easier.

⚠ When rinsing the slide, avoid spraying the water directly on the smear as this may dislodge your specimens.

⚠ Dispose of the specimen slides in a jar of disinfectant after use.

References

Murray, R.G.E., Raymond N. Doetsch and C.F. Robinow. 1994. Page 28 in *Methods for General and Molecular Bacteriology,* edited by Philipp Gerhardt, R.G.E. Murray, Willis A. Wood, and Noel R. Krieg. American Society for Microbiology, Washington, D.C.

Norris, J. R. and Helen Swain. 1971. Chapter II in *Methods in Microbiology, Volume 5A,* edited by J. R. Norris and D. W. Ribbons. Academic Press, Ltd., London.

Bacterial Cellular Structures and Differential Stains

Gram Stain

Photographic Atlas Reference

Page 27

Staining Protocol

1. Prepare and heat-fix a smear of each organism next to one another on the same clean glass slide. Since Gram stains require much practice, you may want to prepare several slides and let them be air drying simultaneously. Then they'll be ready if you need them.
2. Use a sterile toothpick to obtain a sample from your teeth at the gum line. Transfer the sample to a drop of water on a clean glass slide, air dry and heat fix.
3. Follow the basic staining procedure illustrated in Figures 4-1a through 4-1i. We recommend staining the pure cultures first. When your technique is consistent, then stain the tooth sample.
4. Observe using the oil immersion lens. Record your observations of cell morphology and arrangement, dimensions, and Gram reactions in the table provided.

Materials

- Clean glass microscope slides
- Sterile toothpick
- Gram crystal violet
- Gram iodine
- 90% ethanol
- Gram safranin
- Bibulous paper
- Disposable latex gloves
- 18 to 24 hour nutrient agar slant pure cultures of:
  - *Bacillus cereus*
  - *Escherichia coli*

FIGURE 4-1a.

Flood the Smear with Crystal Violet *Place your slide with its smears on the staining rack. Flood the smear with crystal violet and let it stand for one minute. Hold the slide with a slide holder to minimize staining of your hands. Wearing latex gloves is also a good idea.*

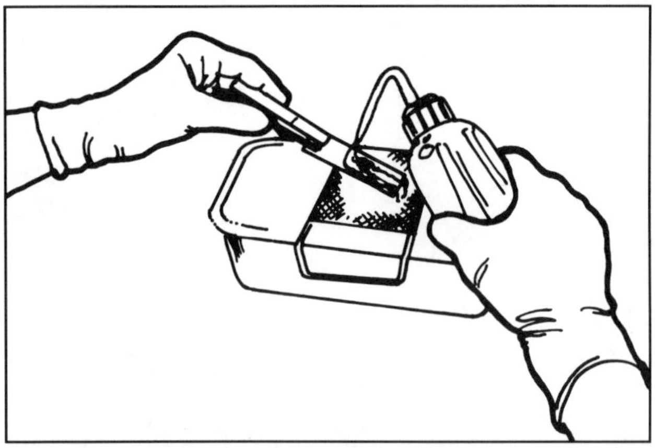

FIGURE 4-1b.

Rinse with Water *Tilt the slide to an angle of 45°. Rinse away excess crystal violet by directing a stream of water toward the top of the slide and allowing the water to run down across the smear's surface. Avoid spraying the water directly on the smear as this may dislodge your specimens.*

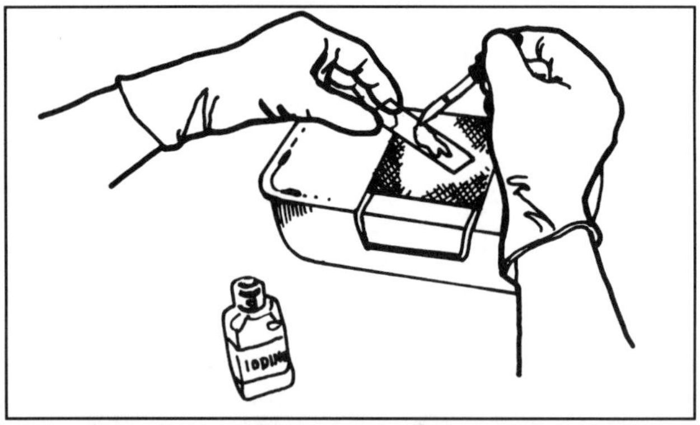

FIGURE 4-1c.

Flood the Smear with Iodine *Cover the smear with iodine solution for at least one minute.*

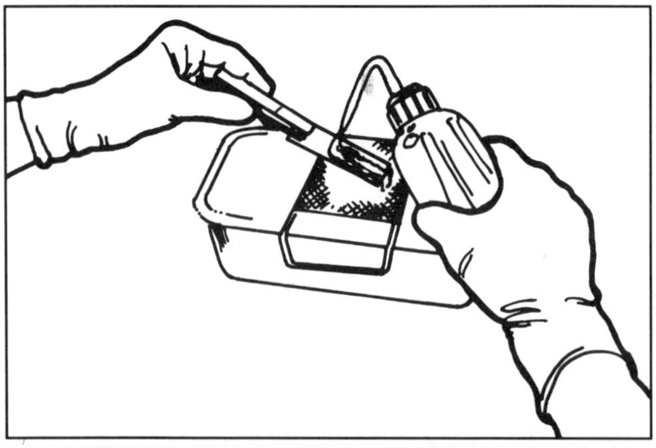

FIGURE 4-1d.

Rinse Again with Water *Gently wash off the excess iodine as in Figure 4-1b.*

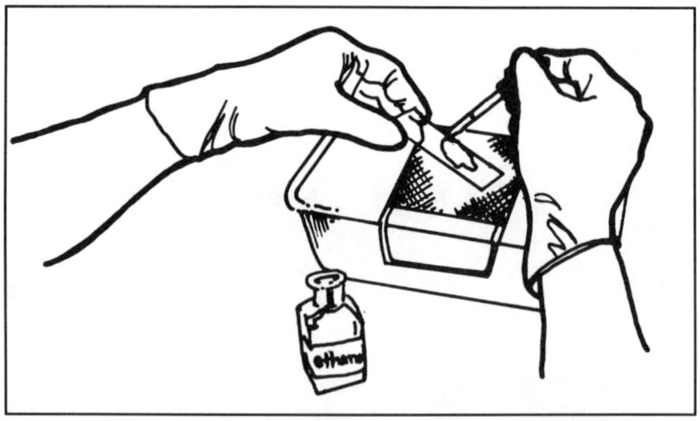

FIGURE 4-1e.

Decolorize *Hold the slide at a 45° angle. Allow the alcohol to run across the smear. Stop decolorizing when the run-off is clear, but no longer than 30 seconds.*

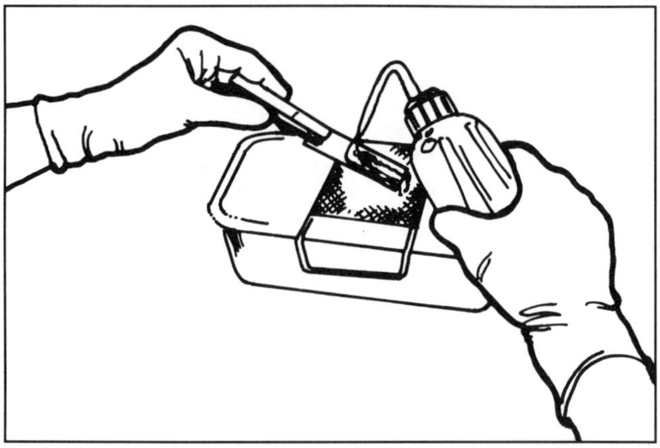

FIGURE 4-1f.

Rinse Again with Water *Immediately rinse off the alcohol with water as in Figure 4-1b. The longer you delay, the greater the likelihood of overdecolorizing.*

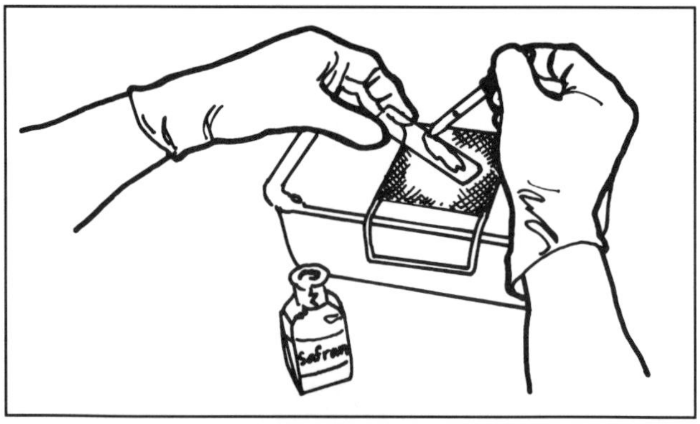

FIGURE 4-1g.

Counterstain with Safranin *Cover the smear with safranin for one minute.*

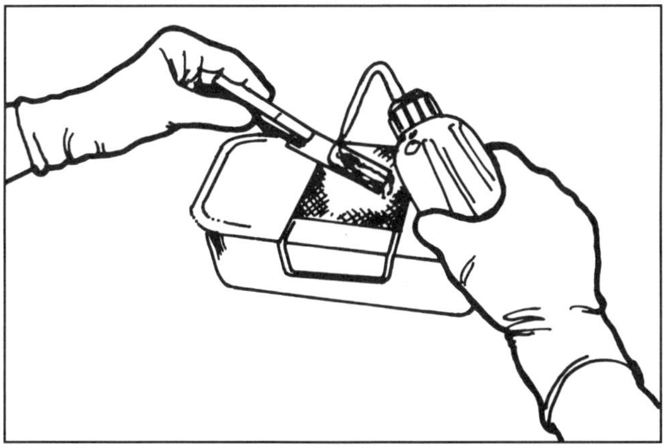

FIGURE 4-1h.

Rinse Again with Water *Gently wash off the excess safranin as in Figure 4-1b.*

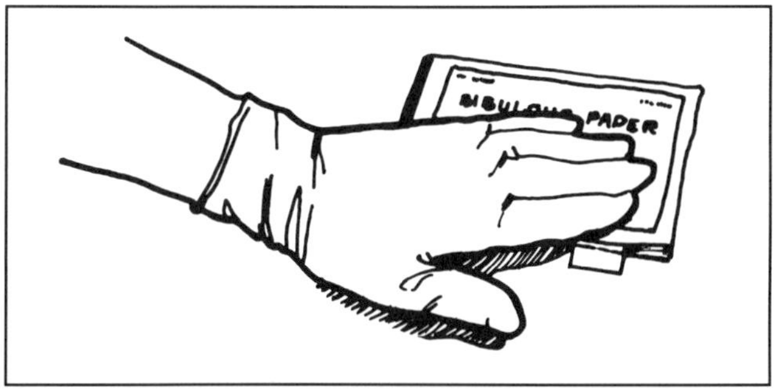

FIGURE 4-1i.

Blot Dry *Blot (do not wipe) the slide dry in a tablet of bibulous paper. Then observe the specimen using the oil immersion lens.*

Precautions

⚠ Some Gram-positive microbes lose their ability to resist decolorization with age. Always use cultures younger than 24 hours old.

⚠ Decolorization is the most critical step in the procedure.

⚠ Strive for preparing smears of uniform thickness. Thick smears risk being underdecolorized, whereas thin smears risk being overdecolorized.

⚠ Until you are confident of your ability to produce consistent and reliable Gram stains, it is a good idea to run controls (known Gram-positive and Gram-negative organisms) next to your specimen (see Fig. 4-3 in the *Atlas*.)

⚠ Dispose of the specimen slides in a jar of disinfectant after use.

References

Chapin, Kimberle. 1995. Chapter 4 in *Manual of Clinical Microbiology, 6th Ed.*, edited by Patrick R. Murray, Ellen Jo Baron, Michael A. Pfaller, Fred C. Tenover, and Robert H. Yolken. American Society for Microbiology, Washington, D.C.

Murray, Patrick R., Ellen Jo Baron, Michael A. Pfaller, Fred C. Tenover, and Robert H. Yolken. 1995. *Manual of Clinical Microbiology, 6th Ed.* American Society for Microbiology, Washington, D.C.

Murray, R.G.E., Raymond N. Doetsch and C.F. Robinow. 1994. Pages 31 and 32 in *Methods for General and Molecular Bacteriology*, edited by Philipp Gerhardt, R.G.E. Murray, Willis A. Wood, and Noel R. Krieg. American Society for Microbiology, Washington, D.C.

Norris, J. R. and Helen Swain. 1971. Chapter II in *Methods in Microbiology, Volume 5A*, edited by J. R. Norris and D. W. Ribbons. Academic Press, Ltd., London.

Power, David A. and Peggy J. McCuen. 1988. Page 4 in *Manual of BBL® Products and Laboratory Procedures, 6th Ed.* Becton Dickinson Microbiology Systems, Cockeysville, MD.

ORGANISM	CELLULAR MORPHOLOGY AND ARRANGEMENT	CELL DIMENSIONS	GRAM REACTION (+/–)
Bacillus cereus			
Escherichia coli			
Tooth sample			
Tooth sample			
Tooth sample			
Tooth sample			

Acid-Fast Stain

Photographic Atlas Reference

Page 29

Staining Protocol

1. Prepare smears of each organism on a clean glass slide, substituting a drop of sheep serum for the drop of water. Air dry, then heat-fix the smears. NOTE: you may make two separate smears right next to one another on the slide or mix the two organisms in one smear.
2. Cover the slide with a strip of bibulous paper and place it on the heating apparatus. The paper strip should be the same size as the slide.
3. Saturate the paper with Ziehl's carbolfuchsin and keep it moist as you steam the slide for 5 minutes (see Figure 4-2). It is advisable to wear latex gloves while staining. A lab coat or apron and eye goggles should protect you from spattering stain in case the slide overheats. Be sure to clean up any spills.
4. After 5 minutes of steaming, remove the paper and rinse off the excess carbolfuchsin with water. Hold the slide on a 45° angle. Aim the water stream above the bacterial smear and allow the water to run across the smear. (See Figure 3-3b on page 19.)
5. Decolorize with acid alcohol until the runoff is clear. Caution! Do not hold the slide with your fingers. This is an acidic solution.
6. Rinse with water as before.
7. Counterstain with methylene blue for one minute. (See Figure 3-3a on page 19.)
8. Rinse with water, then blot dry with bibulous paper.
9. Observe using the oil immersion lens. Record your observations of cell morphology and arrangement, dimensions, and acid-fast reaction in the table provided.

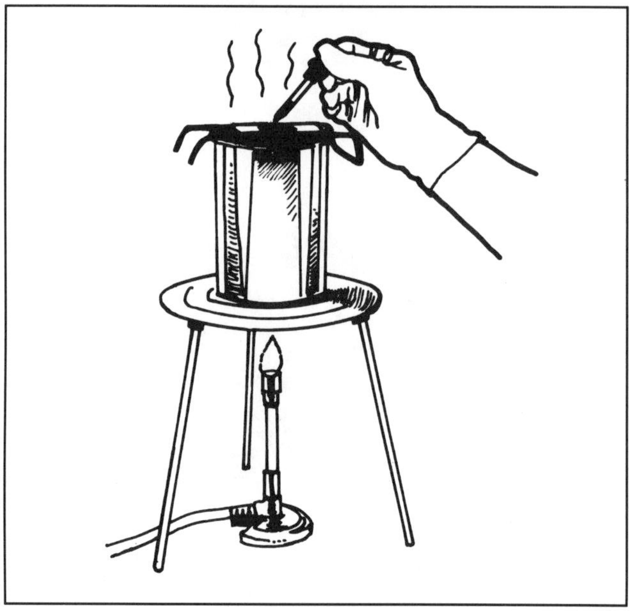

FIGURE 4-2.

Steaming the Slide *Carefully steam the slide to force the carbolfuchsin into acid-fast cells. Do not boil the slide or let it dry out. Keep it moist with stain for the entire five minutes of steaming.*

Materials

- Clean glass microscope slides
- Methylene blue stain
- Ziehl's carbolfuchsin stain
- Acid alcohol (3% HCl in methanol)
- Sheep serum
- Heating apparatus (steam or hot plate)
- Bibulous paper
- Disposable latex gloves
- Lab coat or apron
- Eye goggles
- 18 to 24 hour nutrient agar slant pure cultures of:
  - *Mycobacterium phlei*
  - *Bacillus subtilis*

ORGANISM	CELLULAR MORPHOLOGY AND ARRANGEMENT	CELL DIMENSIONS	ACID-FAST REACTION (+/–)
Bacillus subtilis			
Mycobacterium phlei			

Precautions

- ⚠ Decolorization is not nearly as challenging in this procedure as in the Gram stain, but uniform smear preparation is still necessary to achieve consistent results.
- ⚠ Perform this stain in a well-ventilated area.
- ⚠ Keep the paper moist with stain while steaming it.
- ⚠ Do not boil the stain on the slide. This may cause spattering of stain and crack the slide.
- ⚠ Use the acid alcohol with caution.
- ⚠ Dispose of the specimen slides in a jar of disinfectant after use.

References

Chapin, Kimberle. 1995. Chapter 4 in *Manual of Clinical Microbiology, 6th Ed.*, edited by Patrick R. Murray, Ellen Jo Baron, Michael A. Pfaller, Fred C. Tenover, and Robert H. Yolken. American Society for Microbiology, Washington, D.C.

Murray, R.G.E., Raymond N. Doetsch and C.F. Robinow. 1994. Page 32 in *Methods for General and Molecular Bacteriology*, edited by Philipp Gerhardt, R.G.E. Murray, Willis A. Wood, and Noel R. Krieg. American Society for Microbiology, Washington, D.C.

Norris, J. R. and Helen Swain. 1971. Chapter II in *Methods in Microbiology, Volume 5A*, edited by J. R. Norris and D. W. Ribbons. Academic Press, Ltd., London.

Power, David A. and Peggy J. McCuen. 1988. Page 5 in *Manual of BBL® Products and Laboratory Procedures, 6th Ed.* Becton Dickinson Microbiology Systems, Cockeysville, MD.

Capsule Stain

Photographic Atlas Reference

Page 30

Staining Protocol

Each specimen should be done on a separate slide.

1. Place a small drop of sheep serum at one end of a clean glass slide. Add a drop of Congo red stain, then aseptically transfer a small amount of *F. capsulatum* to the drop and mix. Wearing latex gloves to protect your hands is a good idea. Be sure to flame your inoculating loop after the transfer.
2. Use a clean slide to spread a thin film of the serum/stain/bacteria drop as in the negative stain (see Figures 3-1c through 3-1e). Be sure to dispose of the spreader slide properly.
3. Allow the slide to air dry completely.
4. Flood the slide with Maneval's stain for one minute (see Figure 3-3a).
5. Rinse with water (see Figure 3-3b), then blot dry with bibulous paper.
6. Observe using the oil immersion lens. Record your observations of cell morphology and arrangement, cell dimensions, and presence or absence of a capsule in the table provided.

Materials

- Clean glass slides
- Sheep serum
- Maneval's stain
- Congo red stain
- Bibulous paper
- Latex gloves
- 18 to 24 hour skim milk pure culture of *Flavobacterium capsulatum (Sphingomonas capsulata)*
- 18 to 24 hour nutrient agar slant pure culture of *Bacillus subtilis*

Precautions

⚠ Be sure to sterilize the inoculating loop after the transfer.

⚠ Do not heat-fix the slides. This will produce artifactual white halos around the cells that may be interpreted as capsules.

⚠ Be sure to dispose of the spreader slide properly in disinfectant.

⚠ Dispose of the specimen slide in a jar of disinfectant after use.

References

Murray, R.G.E., Raymond N. Doetsch and C.F. Robinow. 1994. Page 35 in *Methods for General and Molecular Bacteriology*, edited by Philipp Gerhardt, R.G.E. Murray, Willis A. Wood, and Noel R. Krieg. American Society for Microbiology, Washington, D.C.

Norris, J. R. and Helen Swain. 1971. Chapter II in *Methods in Microbiology, Volume 5A*, edited by J. R. Norris and D. W. Ribbons. Academic Press, Ltd, London.

ORGANISM	CELLULAR MORPHOLOGY AND ARRANGEMENT	CELL DIMENSIONS	CAPSULE (+/–)
Bacillus subtilis			
Flavobacterium capsulatum			

Spore Stain

Photographic Atlas Reference

Page 31

Staining Protocol

1. Prepare and heat-fix a smear of each organism on the same slide.
2. Cover the slide with a strip of bibulous paper and place it on the heating apparatus. The paper strip should be the same size as the slide.
3. Saturate the paper with malachite green stain and keep it moist as you steam the slide for 5 minutes (see Figure 4-3). It is advisable to wear latex gloves while staining. A lab coat or apron and eye goggles should protect you from spattering stain in case the slide overheats. Be sure to clean up any spills.
4. After 5 minutes of steaming, remove the paper and gently rinse off the excess malachite green with water. Water also acts as the decolorizing agent in this procedure, so rinse thoroughly. (See Figure 3-3b.)
5. Counterstain with safranin for up to one minute. (See Figure 3-3a.)
6. Rinse gently with water, then blot dry with bibulous paper.

Materials

Clean glass microscope slides
Malachite green stain
Safranin stain
Heating apparatus (steam apparatus or hot plate)
Bibulous paper
Disposable latex gloves
Lab coat or apron
Goggles
18 to 24 hour nutrient agar slant pure culture of *Staphylococcus epidermidis*
5 day nutrient agar slant pure culture of *Bacillus cereus*

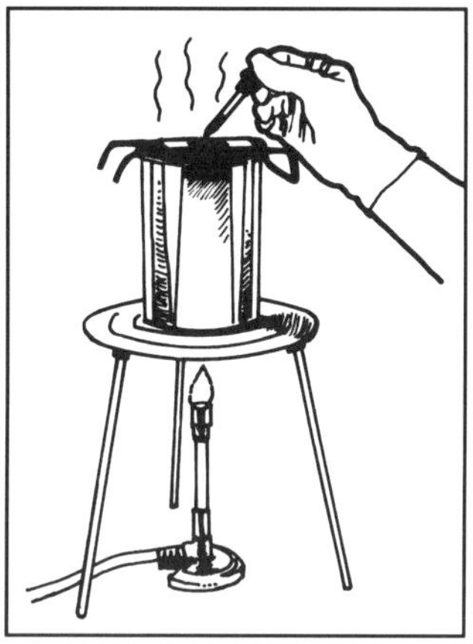

FIGURE 4-3.

Steaming the Slide *Carefully steam the slide to force the malachite green into the spores. Do not boil the slide or let it dry out. Keep it moist with stain for the entire five minutes.*

7. Observe using the oil immersion lens. Record your observations of cell morphology and arrangement, cell dimensions, and spore presence, position and shape in the table provided.

Precautions

⚠ Perform this stain in a well-ventilated area.

⚠ Keep the paper moist with stain while steaming it.

⚠ Do not boil the stain on the slide. This may cause spattering of stain and crack the slide. It may also destroy any vegetative cells present.

⚠ Absence of spores does not necessarily mean the organism *can't* produce them. It may be that the culture is too young for spores to be produced.

⚠ Dispose of the specimen slide in a jar of disinfectant after use.

References

Claus, G. William. 1989. Chapter 9 in *Understanding Microbes — A Laboratory Textbook for Microbiology*. W.H. Freeman and Company, New York, NY.

Murray, R.G.E., Raymond N. Doetsch and C.F. Robinow. 1994. Page 34 in *Methods for General and Molecular Bacteriology*, edited by Philipp Gerhardt, R.G.E. Murray, Willis A. Wood, and Noel R. Krieg. American Society for Microbiology, Washington, D.C.

ORGANISM	CELLULAR MORPHOLOGY AND ARRANGEMENT	CELL DIMENSIONS	SPORES (+/–)	SPORE SHAPE	SPORE POSITION
Bacillus cereus					
Staphylococcus epidermidis					

Flagella Stain

Photographic Atlas Reference

Page 33

Staining Protocol

Each specimen should be done on a separate slide.

1. Clean a slide with a thin solution of household cleanser, such as Bon Ami. Allow the paste to dry in a thin film on the slide, then remove thoroughly with a tissue.
2. Add a few milliliters of nutrient broth to the agar slant with the Pasteur pipette.
3. After a few minutes, decant the broth into a centrifuge tube.
4. Centrifuge (longer is better than faster) until a visible pellet is seen. Decant the supernatant from the tube and resuspend the pellet in 10% (v/v) Formalin. In all of these steps, handle the bacteria gently, as the flagella are fragile and may break off.
5. Transfer a loop of the bacterial suspension to a cleaned slide. Hold the slide on an angle and allow the drop to run down it. Air dry the bacterial film.
6. Draw a rectangle around the film using a china marker
7. Flood the rectangular region with the pararosaniline stain until a golden film with precipitate forms (visible with transmitted light) throughout the preparation. This may take up to 15 minutes.
8. Rinse gently with water, then air dry.
9. Observe using the oil immersion lens. Record your observations of cell morphology, arrangement, and dimensions, and flagellar presence and arrangement in the table provided.
10. Commercially prepared slides are available and provide satisfactory material for observation. Record your observations in the open spaces in the table below.

Materials

Sterile nutrient broth
Pasteur pipettes
Centrifuge and centrifuge tubes
10% Formalin
Wax china marker
Pararosaniline stain
Sterile nutrient broth tube
Commercially prepared slides of motile organisms.
18 to 24 hour nutrient agar slant pure cultures of:
Proteus vulgaris
Staphylococcus epidermidis

Precautions

⚠ To minimize the effects of stain precipitates and other artifacts, only use slides that have been cleaned and degreased.

ORGANISM	CELLULAR MORPHOLOGY AND ARRANGEMENT	CELL DIMENSIONS	FLAGELLA (+/–)	FLAGELLAR ARRANGEMENT
Proteus vulgaris				
Staphylococcus epidermidis				

⚠ Flagella are extremely fragile. Handle the preparations very gently to minimize the inevitable damage.

⚠ Flagella are very small and are not easily seen. Use the oil lens and patiently scan your preparation until you find a field with satisfactory specimens. Even commercially prepared slides may require more than the usual effort to find good examples.

⚠ Dispose of your specimen slide in a jar of disinfectant after use.

References

Iino, Tetsuo and Masatoshi Enomoto. 1971. Chapter IV in *Methods in Microbiology, Volume 5A,* edited by J. R. Norris and D. W. Ribbins. Academic Press, Ltd., London.

Murray, R.G.E., Raymond N. Doetsch and C.F. Robinow. 1994. Page 35 in *Methods for General and Molecular Bacteriology,* edited by Philipp Gerhardt, R.G.E. Murray, Willis A. Wood, and Noel R. Krieg. American Society for Microbiology, Washington, D.C.

Hanging Drop Technique

Photographic Atlas Reference

Page 35

Protocol

1. Follow the procedure illustrated in Figures 4-4a through 4-4d for each specimen.
2. Observe under high dry or oil immersion and record your results in the table below.

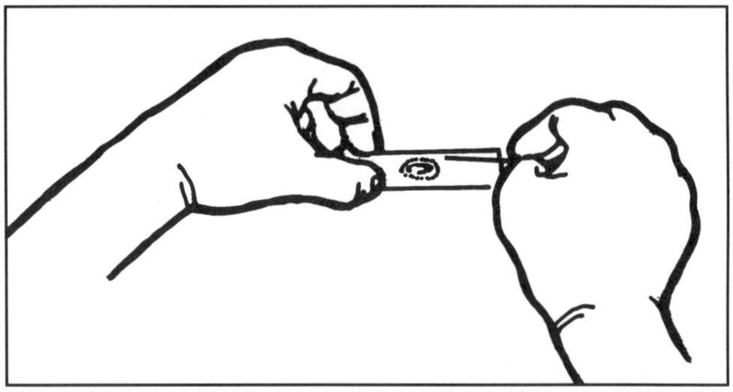

FIGURE 4-4a.

Apply the Petroleum Jelly *Use a toothpick to place a* thin *ring of petroleum jelly around the well of a depression slide.*

Materials

Depression slide and cover glass
Petroleum jelly
Toothpick
18 to 24 hour nutrient agar slant pure cultures of:
Proteus vulgaris
Staphylococcus epidermidis

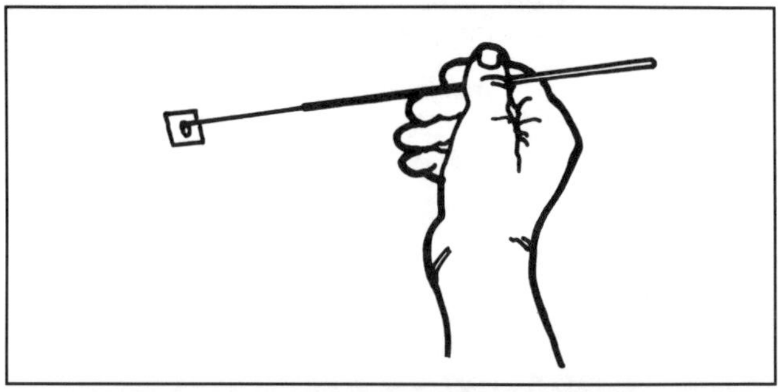

FIGURE 4-4b.

Transfer the Organisms *Use your loop to place a drop of water on a cover glass. Then, aseptically transfer the bacteria to the drop. Do not emulsify or spread out the drop. Flame your loop immediately after the transfer.*

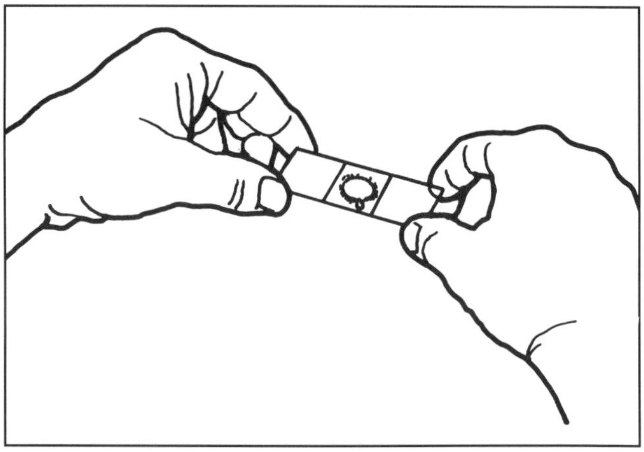

FIGURE 4-4c.

Invert the Slide Over the Cover Glass *Carefully invert the depression slide over the cover glass so the drop is centered in its well. Gently press until the petroleum jelly has created a seal between the slide and cover glass.*

ORGANISM	CELLULAR MORPHOLOGY AND ARRANGEMENT	CELL DIMENSIONS	MOTILITY (+/–)
Proteus vulgaris			
Staphylococcus epidermidis			

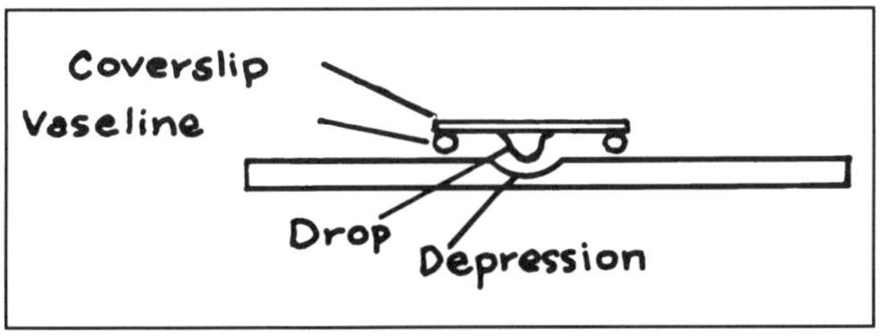

FIGURE 4-4d.

Observe the Slide *Correctly done, the slide should look like this from the side. Place the slide with the cover glass up on the microscope stage and observe under high dry or oil immersion.*

Precautions

⚠ Avoid excessive petroleum jelly on your slide. Petroleum jelly in the well will interfere with your observations. If the layer is too thick, your slide will not fit under the oil immersion lens.

⚠ Be careful not to squash your preparation with the high dry or oil immersion lenses while focusing. There is also the risk of the oil lens pushing the cover glass out of the way as you swing it into place.

⚠ Begin observation at the edge of the drop. This will make focusing easier and may provide a better view of motile microbes if they are aerobic.

⚠ Be sure to distinguish between true motility and Brownian motion due to the kinetic energy of the water molecules crashing into the cells. If the cells are not vibrating at all, they may be stuck on the glass slide.

⚠ Be aware of water currents in the drop that may give the appearance of motility.

⚠ When finished, remove the cover glass and soak it and the depression slide in a jar of disinfectant. After 15 minutes, remove them, rinse with water, and clean with ethanol to remove the petroleum jelly. Rinse with water and dry.

References

Iino, Tetsuo and Masatoshi Enomoto. 1969. Chapter IV in *Methods in Microbiology, Volume 1,* edited by J. R. Norris and D. W. Ribbins. Academic Press, Ltd., London.

Murray, R.G.E., Raymond N. Doetsch and C.F. Robinow. 1994. Page 26 in *Methods for General and Molecular Bacteriology,* edited by Philipp Gerhardt, R.G.E. Murray, Willis A. Wood, and Noel R. Krieg. American Society for Microbiology, Washington, D.C.

Quesnel, Louis B. 1969. Chapter X in *Methods in Microbiology, Volume 1,* edited by J. R. Norris and D. W. Ribbins. Academic Press, Ltd., London.

Miscellaneous Structures

Photographic Atlas Reference

Page 36

Materials

Commercially prepared slides stained for bacterial nucleoplasm

Commercially prepared slides stained for poly-ß-hydroxybutyrate (PHB) granules

Procedure

Observe the slides under oil immersion and record your results below.

STAIN	APPEARANCE
nucleoplasm	
PHB granules	

SECTION FIVE

Differential Tests

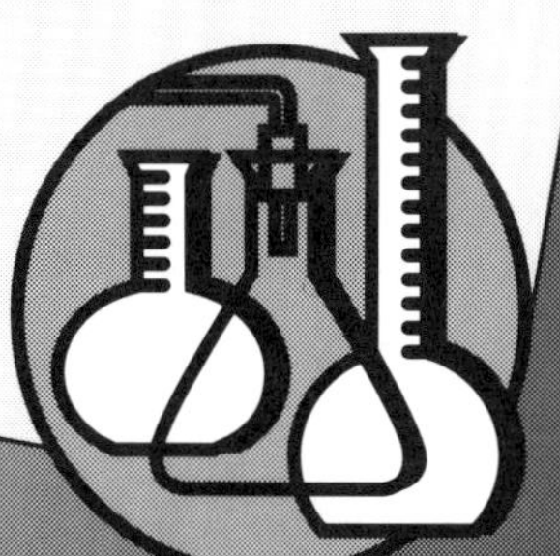

Bile Esculin Agar

Photographic Atlas Reference

Page 37

Recipe

Bile Esculin Agar

Beef extract	3.0 g
Peptone	5.0 g
Oxgall	40.0 g
Esculin	1.0 g
Ferric citrate	0.5 g
Agar	15.0 g
Distilled or deionized water	1.0 L

final pH = 6.6 ± 0.2 at 25°C

Materials

Three bile esculin agar slants
18 to 24 hour pure cultures of:

Proteus mirabilis
Streptococcus faecalis

Medium Preparation

1. Suspend the ingredients in one liter of distilled or deionized water, mix well and boil to dissolve completely.
2. Dispense 7.0 mL volumes into test tubes and cap loosely.
3. Sterilize in the autoclave at 15 lbs. pressure (121°C) for 15 minutes.
4. Remove from the autoclave, slant and allow to cool to room temperature.

Test Protocol

1. Inoculate two slants with the test organisms and leave the third slant uninoculated as a control.
2. Label the slants with the organisms' names, your name and the date.
3. Incubate the slants with the uninoculated control at 35°C for up to 72 hours.
4. After incubation, observe all the tubes for blackening of the agar. Any blackening is scored as positive.
5. Record your results in the table on the next page.

Precaution

△ Incubate for the full 72 hours to reduce the chance of false negatives.

References

DIFCO Laboratories. 1984. Page 129 in *DIFCO Manual, 10th Ed.* DIFCO Laboratories, Detroit, MI.

Lányi, B. 1987. Page 56 in *Methods in Microbiology, Vol. 19*, edited by R. R. Colwell and R. Grigorova, Academic Press Inc., New York.

MacFaddin, Jean F. 1980. Page 4 in *Biochemical Tests for Identification of Medical Bacteria, 2nd Ed.* Williams & Wilkins, Baltimore, MD.

Power, David A. and Peggy J. McCuen. 1988. Page 113 in *Manual of BBL® Products and Laboratory Procedures, 6th Ed.* Becton Dickinson Microbiology Systems, Cockeysville, MD.

ORGANISM	RESULT +/-
Proteus mirabilis	
Streptococcus faecalis	

Blood Agar

Photographic Atlas Reference

Page 39

Recipe

Blood Agar

Infusion from beef heart (solids)	2.0 g
Pancreatic digest of casein	13.0 g
Sodium chloride	5.0 g
Yeast extract	5.0 g
Agar	15.0 g
Defibrinated sheep blood	50.0 mL
Distilled or deionized water	1.0 L

final pH = 7.3 ± 0.2 at 25°C

Medium Preparation

(See Precautions below.)

1. Suspend, mix and boil the dry ingredients in one liter distilled or deionized water. This is blood agar base.
2. Cover loosely and sterilize in the autoclave at 15 lbs. pressure (121°C) for 15 minutes.
3. Remove from the autoclave and cool to 45°C.
4. Aseptically add the sterile, room temperature sheep blood to the blood agar base and mix well.
5. Pour into sterile Petri dishes and allow to cool to room temperature.

Materials

One blood agar plate
18 to 24 hour pure cultures of:
Proteus mirabilis
Staphylococcus aureus
Staphylococcus epidermidis

Test Protocol

1. Using a marking pen, divide the blood agar plate into four equal sectors. Be sure to mark the bottom of the plate.
2. Spot inoculate three sectors with the test organisms leaving the fourth sector as a control. (Spot inoculations are for demonstration only. This medium is intended to be streaked for isolation.)
3. Label the plate with the organisms' names, your name and the date.
4. Invert and incubate the plate aerobically at 35°C for 24 hours.
5. Observe for color and clearing around the growth.
6. Record your results in the table below.

Precautions

⚠ We recommend buying commercially prepared 5% Sheep Blood Agar plates.

⚠ Blood agar plates must be stored in the refrigerator.

References

DIFCO Laboratories. 1984. Page 139 in *DIFCO Manual, 10th Ed.* DIFCO Laboratories, Detroit, MI.

Krieg, Noel R. 1994. Page 619 in *Methods for General and Molecular Bacteriology*, edited by Philipp Gerhardt, R. G. E. Murray, Willis A. Wood and Noel R. Krieg, American Society for Microbiology, Washington, DC.

Power, David A. and Peggy J. McCuen. 1988. Page 115 in *Manual of BBL® Products and Laboratory Procedures, 6th Ed.* Becton Dickinson Microbiology Systems, Cockeysville, MD.

ORGANISM	RESULT
Proteus mirabilis	
Staphylococcus aureus	
Staphylococcus epidermidis	

β = Complete hemolysis; α = partial hemolysis; γ = no hemolysis

Catalase Test

Photographic Atlas Reference

Page 40

Recipe

Nutrient Agar

Beef extract	3.0 g
Peptone	5.0 g
Agar	15.0 g
Distilled or deionized water	1.0 L

final pH = 6.8 ± 0.2 at 25°C

Medium Preparation

1. Suspend, mix and boil the ingredients in one liter of distilled or deionized water to dissolve completely.
2. Dispense 7.0 mL volumes into test tubes and cap loosely.
3. Sterilize in the autoclave at 15 lbs. pressure (121°C) for 15 minutes.
4. Remove from the autoclave, slant and allow to cool to room temperature.

Test Protocol

1. Streak inoculate two slants with the test organisms and leave the third slant uninoculated as a control.
2. Label the slants with the organisms' names, your name and the date.
3. Incubate the slants aerobically with the uninoculated control at 35°C for 24 hours.
4. After incubation, perform the tests as follows:

   Slide test

   a. Transfer a large amount of growth to a microscope slide.
   b. Apply 1 or 2 drops of hydrogen peroxide directly to the bacteria on the slide.
   c. Observe for the formation of bubbles.

   Slant test

   a. Place several drops of hydrogen peroxide directly into the slant containing the fresh culture.
   b. Observe for the formation of bubbles.
5. Record your results in the table below.

Materials

Three nutrient agar slants
Hydrogen peroxide (3%)
Microscope slides
18 to 24 hour pure cultures of:
Staphylococcus epidermidis
Streptococcus faecalis

Precautions

⚠ When performing the slide test, place the bacteria on the slide first and then add the hydrogen peroxide. Placing the inoculating needle in the hydrogen peroxide may catalyze a false positive reaction.

⚠ Do not use blood agar as the growth medium; catalase in the blood will result in false positive reactions.

⚠ To reduce the chance of false negative results, use only fresh cultures for this test (24 hours or less).

References

Collins, C. H., Patricia M. Lyne, J. M. Grange. 1995. Page 110 in *Collins and Lyne's Microbiological Methods, 7th Ed.*, Butterworth-Heinemann, UK.

DIFCO Laboratories. 1984. Page 619 in *DIFCO Manual, 10th Ed.* DIFCO Laboratories, Detroit, MI.

Lányi, B. 1987. Page 20 in *Methods in Microbiology, Vol. 19*, edited by R. R. Colwell and R. Grigorova, Academic Press Inc., New York.

MacFaddin, Jean F. 1980. Page 51 in *Biochemical Tests for Identification of Medical Bacteria, 2nd Ed.*, Williams & Wilkins, Baltimore, MD.

Smibert, Robert M. and Noel R. Krieg. 1994. Page 614 in *Methods for General and Molecular Bacteriology*, edited by Philipp Gerhardt, R. G. E. Murray, Willis A. Wood and Noel R. Krieg, American Society for Microbiology, Washington, DC.

ORGANISM	RESULT +/–
Staphylococcus epidermidis	
Streptococcus faecalis	

Citrate Utilization Agar

Photographic Atlas Reference

Page 41

Recipe

Simmons Citrate Agar

Ammonium dihydrogen phosphate	1.0 g
Dipotassium phosphate	1.0 g
Sodium chloride	5.0 g
Sodium citrate	2.0 g
Magnesium sulfate	0.2 g
Agar	15.0 g
Bromthymol blue	0.08 g
Distilled or deionized water	1.0 L

final pH = 6.9 ± 0.2 at 25°C

Medium Preparation

1. Suspend the ingredients in one liter distilled or deionized water, mix well and boil to dissolve completely.
2. Dispense 7.0 mL portions into test tubes and cap loosely.
3. Sterilize in the autoclave at 15 lbs pressure (121°C) for 15 minutes.
4. Remove from the autoclave, slant and allow to cool to room temperature.

Materials

Three citrate agar slants
18 to 24 hour pure cultures of:

Enterobacter aerogenes
Escherichia coli

Test Protocol

1. Using a light inoculum, streak inoculate two slants with the test organisms. Leave the third slant uninoculated as a control.
2. Label the slants with the organisms' names, your name and the date.
3. Incubate the tubes aerobically with the uninoculated control at 35°C for up to 4 days.
4. Observe the tubes for color changes.
5. Record your results in the table below.

Precaution

⚠ Formation of blue color in any portion of the slant is considered a positive reaction.

References

Collins, C. H., Patricia M. Lyne, J. M. Grange. 1995. Page 111 in *Collins and Lyne's Microbiological Methods, 7th Ed.* Butterworth-Heinemann, UK.

DIFCO Laboratories. 1984. Page 864 in *DIFCO Manual, 10th Ed.* DIFCO Laboratories, Detroit, MI.

Power, David A. and Peggy J. McCuen. 1988. Page 246 in *Manual of BBL® Products and Laboratory Procedures, 6th Ed.* Becton Dickinson Microbiology Systems, Cockeysville, MD.

Smibert, Robert M. and Noel R. Krieg. 1994. Page 614 in *Methods for General and Molecular Bacteriology*, edited by Philipp Gerhardt, R. G. E. Murray, Willis A. Wood and Noel R. Krieg, American Society for Microbiology, Washington, DC.

ORGANISM	RESULT +/–
Enterobacter aerogenes	
Escherichia coli	

Coagulase Test

Photographic Atlas Reference

Page 42

Medium Preparation

1. Add the recommended volume of sterile deionized or distilled water to the dehydrated medium.
2. Mix well.
3. Aseptically transfer 0.5 mL volumes to small sterile test tubes.

Test Protocol

1. Inoculate two plasma tubes with the test organisms and leave a third tube uninoculated as a control.
2. Label the tubes with the organisms' names, your name and the date.
3. Incubate the tubes with the uninoculated control, at 35°C for up to 24 hours, checking for coagulation every 30 minutes for the first 2 to 4 hours.
4. Record your results in the table below.

Materials

Commercially prepared dehydrated rabbit plasma
Sterile distilled or deionized water
Small sterile test tubes (12 mm x 75 mm)
Sterile 1 mL pipettes
18 to 24 hour pure cultures of:
Staphylococcus aureus
Staphylococcus epidermidis

Precaution

⚠ To reduce the chance of obtaining false negative results for this test, use only fresh cultures (24 hours or less).

⚠ Incubation longer than 24 hours may lead to false negative results in coagulase positive organisms that produce clot digesting enzymes.

References

Collins, C. H., Patricia M. Lyne, J. M. Grange. 1995. Page 111 in *Collins and Lyne's Microbiological Methods, 7th Ed.* Butterworth-Heinemann, UK.

DIFCO Laboratories. 1984. Page 232 in *DIFCO Manual, 10th Ed.* DIFCO Laboratories, Detroit, MI.

Holt, John G. (Editor). 1994. *Bergey's Manual of Determinative Bacteriology, 9th Ed.* Williams and Wilkins, Baltimore, MD.

Lányi, B. 1987. Page 62 in *Methods in Microbiology, Vol. 19*, edited by R. R. Colwell and R. Grigorova, Academic Press Inc., New York.

MacFaddin, Jean F. 1980. Page 64 in *Biochemical Tests for Identification of Medical Bacteria, 2nd Ed.* Williams & Wilkins, Baltimore, MD.

ORGANISM	RESULT +/-
Staphylococcus aureus	
Staphylococcus epidermidis	

Decarboxylase Medium

Photographic Atlas Reference

Page 43

Recipe

Moeller (Møller) Decarboxylase Medium

Peptone	5.0 g
Beef extract	5.0 g
Glucose (dextrose)	0.5 g
Bromcresol purple	0.01 g
Cresol red	0.005 g
Pyridoxal	0.005 g
L-Lysine, L-Ornithine, or L-Arginine	10.0 g
Distilled or deionized water	1.0 L

final pH = 6.0 ± 0.2 at 25°C

Medium Preparation

1. Suspend and heat the ingredients in one liter of distilled or deionized water until completely dissolved. (Use only one of the listed L-amino acids.)
2. Adjust pH by adding NaOH if necessary.
3. Dispense 7.0 mL volumes into test tubes and cap loosely.
4. Sterilize in the autoclave at 15 lbs. pressure (121°C) for 10 minutes.
5. Remove from the autoclave and allow to cool to room temperature.

Materials

Four decarboxylase tubes for each amino acid being tested
Sterile mineral oil
18 to 24 hour pure cultures of:
Alcaligenes faecalis
Enterobacter aerogenes
Pseudomonas aeruginosa

Test Protocol

1. Lightly inoculate the media with the test organisms, leaving one tube uninoculated as a control.
2. Overlay all tubes (including the uninoculated controls) with 2 to 3 mL sterile mineral oil. (See Figure 5-1.)
3. Label the tubes with the organisms' names, your name and the date.
4. Incubate the tubes aerobically with the uninoculated control at 35°C for up to one week. More time may be necessary for weak reactions.
5. Examine tubes for characteristic color changes.
6. Record your results in the table below.

Precautions

⚠ To reduce the chance of obtaining false negative results, avoid reading the tubes too early.

⚠ Be sure to compare the incubated tubes to an uninoculated control since the color changes may be subtle. A color change to yellow indicates fermentation, *not* decarboxylase activity.

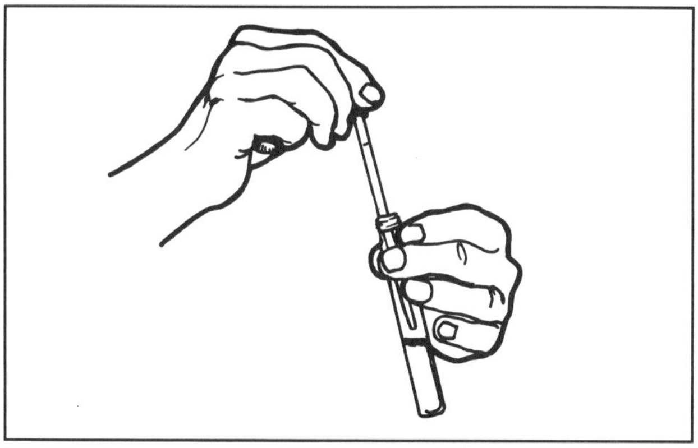

FIGURE 5-1.

Adding the Mineral Oil Layer *Tip the tube slightly to one side and gently add 2 to 3 mL mineral oil. Be sure to use a sterile pipette for* each *tube.*

ORGANISM	L-LYSINE +/−	L-ORNITHINE +/−	L-ARGININE +/−
Alcaligenes faecalis			
Enterobacter aerogenes			
Pseudomonas aeruginosa			

References

Collins, C. H., Patricia M. Lyne, J. M. Grange. 1995. Page 111 in *Collins and Lyne's Microbiological Methods, 7th Ed.* Butterworth-Heinemann, UK.

DIFCO Laboratories. 1984. Page 268 in *DIFCO Manual, 10th Ed.* DIFCO Laboratories, Detroit, MI.

Lányi, B. 1987. Page 29 in *Methods in Microbiology, Vol. 19*, edited by R. R. Colwell and R. Grigorova, Academic Press Inc., New York.

MacFaddin, Jean F. 1980. Page 78 in *Biochemical Tests for Identification of Medical Bacteria, 2nd Ed.* Williams & Wilkins, Baltimore, MD.

DNase Test Agar

Photographic Atlas Reference

Page 45

Recipe

DNase Test Agar with Methyl Green

Tryptose	20.0 g
Deoxyribonucleic acid	2.0 g
Sodium chloride	5.0 g
Agar	15.0 g
Methyl green	0.05 g
Distilled or deionized water	1.0 L

final pH = 7.3 ± 0.2 at 25°C

Medium Preparation

1. Suspend, mix and boil the ingredients in one liter of distilled or deionized water until completely dissolved.
2. Cover loosely and sterilize in the autoclave at 15 lbs. pressure (121°C) for 15 minutes.
3. Aseptically pour into sterile Petri dishes (15 mL/plate) and allow to cool to room temperature.

Test Protocol

1. Using a marking pen, divide the DNase test agar plate into three equal sectors. Be sure to mark the bottom of the plate.

Materials

One DNase test agar plate
18 to 24 hour cultures of:
Enterobacter aerogenes
Serratia marcescens

2. Spot inoculate two sectors with the test organisms and leave the third sector as a control. (Spot inoculations are for demonstration only. These plates are designed to be streaked for isolation.)
3. Label the plate with the organisms' names, your name and the date.
4. Invert the plate and incubate it aerobically at 35°C for 24 hours.
5. Examine the plates for clearing around the bacterial growth. (See Precautions below.)
6. Record your results in the table below.

Precaution

⚠ DNase activity usually occurs quickly. Be sure to read this test no more than 24 hours after inoculating to prevent excessive clearing.

References

Collins, C. H., Patricia M. Lyne, J. M. Grange. 1995. Page 114 in *Collins and Lyne's Microbiological Methods, 7th Ed.* Butterworth-Heinemann, UK.

DIFCO Laboratories. 1984. Page 263 in *DIFCO Manual, 10th Ed.* DIFCO Laboratories, Detroit, MI.

Lányi, B. 1987. Page 33 in *Methods in Microbiology, Vol. 19*, edited by R. R. Colwell and R. Grigorova, Academic Press Inc., New York.

MacFaddin, Jean F. 1980. Page 94 in *Biochemical Tests for Identification of Medical Bacteria, 2nd Ed.* Williams & Wilkins, Baltimore, MD.

Power, David A. and Peggy J. McCuen. 1988. Page 147 in *Manual of BBL® Products and Laboratory Procedures, 6th Ed.* Becton Dickinson Microbiology Systems, Cockeysville, MD.

ORGANISM	RESULT +/–
Enterobacter aerogenes	
Serratia marcescens	

Enterotube® II

Photographic Atlas Reference

Page 47

Test Protocol

1. Remove both end caps from the Enterotube® II. Remove the blue cap first and then the white cap, being careful not to contaminate the sterile wire tip. Place the caps open end down on a disinfectant soaked paper towel.
2. Using the tip of the wire, aseptically remove a large amount of growth from one of the plated colonies (Figure 5-2a). Try not to remove any of the agar with it.
3. Grasp the looped end of the wire and while turning, gently pull it through all the segments of the tube (Figure 5-2b). It is not necessary to completely remove the wire but be certain that all the compartments get inoculated.

Materials

Becton Dickinson Microbiology Systems' Enterotube® II Identification System for Enterobacteriaceae

Materials Provided in the Kit

- Enterotube® II
- Results pad
- Color reaction chart

Materials Required But Not Provided in the Kit

- The Computer Coding and Identification System (Identification booklet from Becton Dickinson Microbiology Systems)
- Kovac's reagent*
- Voges-Proskauer reagents**
- Any bacterial culture from the family Enterobacteriaceae growing on a nutrient agar plate. For a class project, a variety of cultures provided as "unknowns" works best.
- Syringe

*The recipe for Kovac's reagent can be found in the Sulfur-Indole-Motility unit in this section.

**The recipe for Voges-Proskauer reagents differs from that described in the Methyl Red and Voges-Proskauer unit. This test requires 0.3% creatine in 20% KOH and 5% α-naphthol in absolute ethanol.

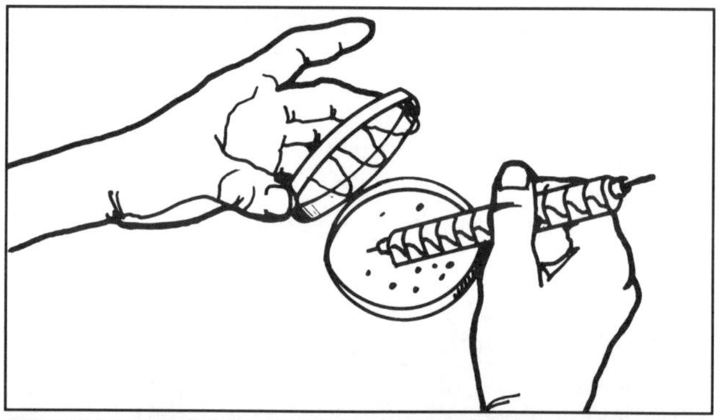

FIGURE 5-2a.

Harvest Growth *Using the sterile tip of the wire, aseptically remove growth from a colony on the agar surface. Do not dig into the agar. The inoculum should be large enough to be visible.*

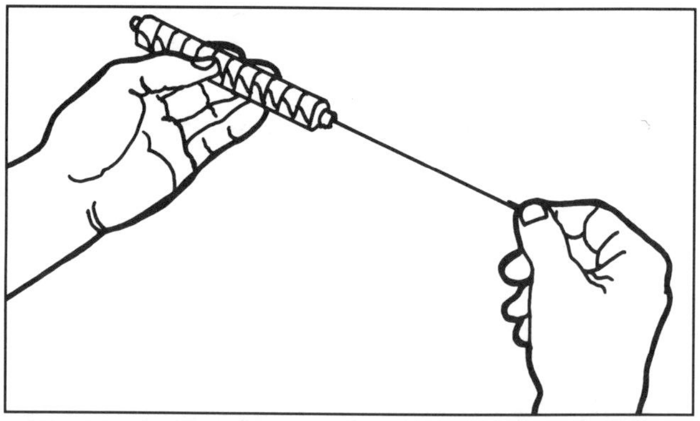

FIGURE 5-2b.

Pull Wire Through *Loosen the wire by turning it slightly. While continuing to rotate it withdraw the wire until its tip is inside the last compartment (glucose).*

4. Using the same turning motion as described above, slide the wire back into the tube (Figure 5-2c). Push it in until the tip of the wire is inside the citrate compartment and the notch in the wire lines up with the opposite end of the tube.
5. Bend the wire at the notch until it breaks off (Figure 5-2d).
6. Locate the air inlets on the side of the tube opposite the Enterotube® II label. Using the removed piece of inoculating wire, puncture the plastic membrane in the adonitol, lactose, arabinose, sorbitol, Voges-Proskauer, dulcitol/PA, urea and citrate compartments (Figure 5-2e). These openings will create the necessary aerobic conditions for growth.

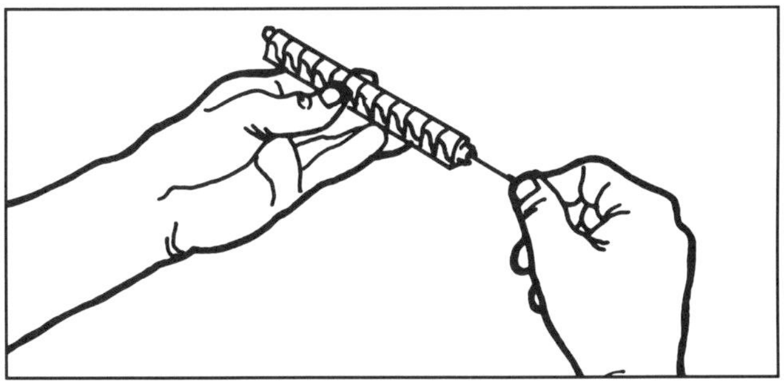

FIGURE 5-2c.

Reinsert Wire *Slide the wire back into the Enterotube® II until you can see the tip inside the citrate compartment. The notch in the wire should be lined up with the end of the tube nearest the glucose compartment.*

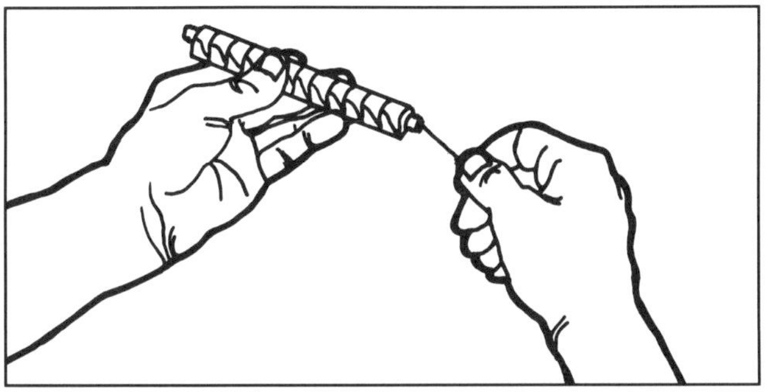

FIGURE 5-2d.

Break Wire *Bend the wire until it breaks off at the notch.*

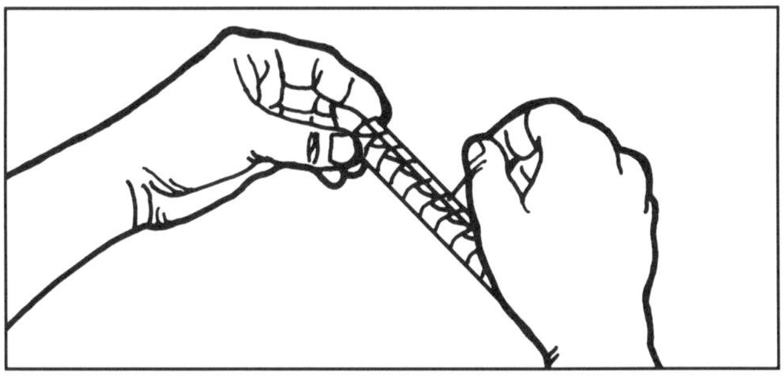

FIGURE 5-2e.

Puncture Air Inlets *Using the broken wire, puncture the plastic covering the eight air inlets on the back side of the tube.*

7. Discard the wire in the appropriate container and replace the Enterotube® II caps.
8. Incubate the tube lying flat at 35°C for 18 to 24 hours.
9. Examine the tube and record the results on the Results Pad.
10. *After* reading and recording all test results place the tube in a rack with the glucose compartment on the bottom. Puncture the plastic membrane of the H_2S/Indole compartment with the needle and syringe containing Kovac's reagent. Add one or two drops of the reagent to the compartment. (Figure 5-2f).
11. Observe the compartment for the formation of a red color within ten seconds.
12. Record the results on the Results Pad.

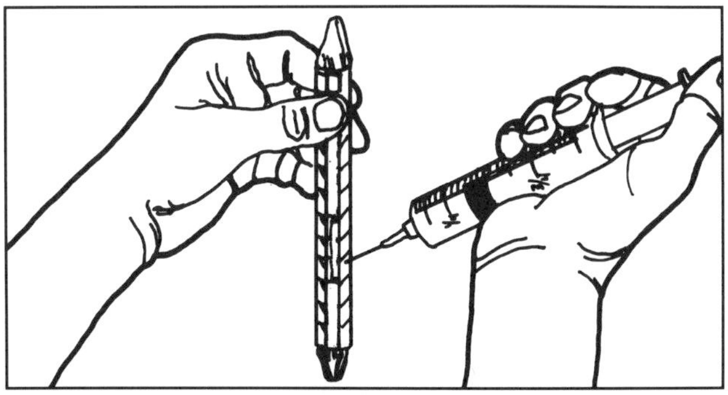

FIGURE 5-2f.

Add Reagents *Use a needle and syringe to add the reagents through the plastic film on the flat side of the tube. Do this with the tube in a test tube rack and the glucose end pointing down. The indole test should be done only after other tests have been read. Add the Voges-Proskauer reagents only if directed to do so by the CCIS.*

13. Interpret the results using the Computer Coding and Identification System (CCIS).
14. If directed to do so by the CCIS, run a confirmatory Voges-Proskauer test. Do this in the same manner as you did the indole test but add the reagents as follows:
    a. Add two drops of 20% potassium hydroxide solution with 0.3% creatine.
    b. Add three drops of 5% alpha-naphthol in absolute ethanol.
15. Observe for the formation of red color within twenty minutes.
16. Discard the tube in the appropriate autoclave container.

Precaution

⚠ When puncturing the air inlets be careful not to perforate the plastic film covering the flat side of the tube.

Reference

Package insert from Enterotube® II Identification System for Enterobacteriaceae.

Entertube II® test procedure courtesy of Becton Dickinson and Company.

Gelatin Liquefaction Test (Nutrient Gelatin)

Photographic Atlas Reference

Page 49

Recipe

Nutrient Gelatin

Beef extract	3.0 g
Peptone	5.0 g
Gelatin	120.0 g
Distilled or deionized water	1.0 L

final pH = 6.8 ± 0.2 at 25°C

Medium Preparation

1. Slowly add the ingredients to one liter of distilled or deionized water while stirring.
2. Warm to 50°C and maintain temperature until completely dissolved.
3. Dispense 7.0 mL volumes into test tubes and cap loosely.
4. Sterilize in the autoclave at 15 lbs. pressure (121°C) for 15 minutes.
5. Remove from the autoclave immediately and allow to cool to room temperature in the upright position.

Test Protocol

1. Stab inoculate two nutrient gelatin tubes with the test organisms. Leave the third tube uninoculated as a control.
2. Label the tubes with the organisms' names, your name and the date.
3. Incubate the tubes aerobically along with the uninoculated control tube at room temperature for up to one week.
4. Examine the control tube. If the control tube is solid, the test can be read. If the control tube has become liquefied due to the temperature, all tubes must be refrigerated or otherwise cooled until the control is resolidified.
5. Examine the tubes for gelatin liquefaction.
6. Record your results in the table below.

Materials

Three nutrient gelatin stab tubes
18 to 24 hour pure cultures of:
Bacillus subtilis
Escherichia coli

Precautions

⚠ It is important not to overheat this medium during preparation.

⚠ Nutrient gelatin can be incubated at 35°C but it will melt at this temperature. Therefore, incubate the inoculated media in parallel with an uninoculated control and, as described above, cool the media before reading.

⚠ Some organisms take up to 6 weeks to liquefy the gelatin. When incubating for such long time periods, take care to minimize evaporation of the liquefied medium.

References

Collins, C. H., Patricia M. Lyne, J. M. Grange. 1995. Page 112 in *Collins and Lyne's Microbiological Methods, 7th Ed.* Butterworth-Heinemann, UK.

DIFCO Laboratories. 1984. Page 35 in *DIFCO Manual, 10th Ed.* DIFCO Laboratories, Detroit, MI.

Lányi, B. 1987. Page 44 in *Methods in Microbiology, Vol. 19*, edited by R. R. Colwell and R. Grigorova, Academic Press Inc., New York.

MacFaddin, Jean F. 1980. Page 128 in *Biochemical Tests for Identification of Medical Bacteria, 2nd Ed.* Williams & Wilkins, Baltimore, MD.

Power, David A. and Peggy J. McCuen. 1988. Page 215 in *Manual of BBL® Products and Laboratory Procedures, 6th Ed.* Becton Dickinson Microbiology Systems, Cockeysville, MD.

Smibert, Robert M. and Noel R. Krieg. 1994. Page 617 in *Methods for General and Molecular Bacteriology*, edited by Philipp Gerhardt, R. G. E. Murray, Willis A. Wood and Noel R. Krieg, American Society for Microbiology, Washington, DC.

ORGANISM	RESULT +/–
Bacillus subtilis	
Escherichia coli	

Kligler's Iron Agar

Photographic Atlas Reference

Page 50

Recipe

Kligler's Iron Agar

Beef extract	3.0 g
Yeast extract	3.0 g
Peptone	15.0 g
Proteose peptone	5.0 g
Lactose	10.0 g
Dextrose (glucose)	1.0 g
Ferrous sulfate	0.2 g
Sodium chloride	5.0 g
Sodium thiosulfate	0.3 g
Agar	12.0 g
Phenol red	0.024 g
Distilled or deionized water	1.0 L

final pH = 7.4 ± 0.2 at 25°C

Medium Preparation

1. Suspend, mix and boil the ingredients in one liter of distilled or deionized water until completely dissolved.
2. Transfer 7.0 mL portions to test tubes and cap loosely.
3. Sterilize in the autoclave at 15 lbs. pressure (121°C) for 15 minutes.
4. Remove from the autoclave and slant so that the butt portion of the tube remains approximately 3 cm deep.
5. Allow to cool to room temperature.

Materials

Five KIA slants
18 to 24 hour pure cultures on solid media:

Alcaligenes faecalis
Citrobacter diversus
Citrobacter freundii
Proteus vulgaris

Test Protocol

1. Stab inoculate four KIA slants with the test organisms. Using a heavy inoculum stab the agar butt then, as you withdraw the needle, streak the entire slant. (See Precautions below.)
2. Label the slants with the organisms' names, your name and the date.
3. Incubate the slants aerobically with the uninoculated control at 35°C for no less than 18 hours or more than 24 hours.
4. Examine the tubes for characteristic color changes and gas production.
5. Record your results in the table below.

Precautions

- ⚠ The inoculation of butt and slant can be done in either order but both methods require a heavy inoculum.
- ⚠ Time is critical in this test; it must be read between 18 and 24 hours after inoculation.
- ⚠ The medium should be fresh. If it is more than 24 hours old, melt it in boiling water and reslant.
- ⚠ If gas is produced it will appear as bubbles or fissures in the medium or will force the entire agar butt off the bottom of the tube.

ORGANISM	RESULT
Alcaligenes faecalis	
Citrobacter diversus	
Citrobacter freundii	
Proteus vulgaris	

A = acid production; Alk = alkaline reaction; H_2S = H_2S production; G = gas production

References

DIFCO Laboratories. 1984. Page 485 in *DIFCO Manual, 10th Ed.* DIFCO Laboratories, Detroit, MI.

Lányi, B. 1987. Page 44 in *Methods in Microbiology, Vol. 19*, edited by R. R. Colwell and R. Grigorova, Academic Press Inc., New York.

MacFaddin, Jean F. 1980. Page 183 in *Biochemical Tests for Identification of Medical Bacteria, 2nd Ed.* Williams & Wilkins, Baltimore, MD.

Power, David A. and Peggy J. McCuen. 1988. Page 171 in *Manual of BBL® Products and Laboratory Procedures, 6th Ed.* Becton Dickinson Microbiology Systems, Cockeysville, MD.

Litmus Milk Medium

Photographic Atlas Reference

Page 52

Recipe

Litmus Milk Medium

Skim milk	100.0 g
Azolitmin	0.5 g
Sodium sulfite	0.5 g
Distilled or deionized water	1.0 L

final pH = 6.5 ± 0.2 at 25°C

Medium Preparation

1. Suspend and mix the ingredients in one liter of deionized or distilled water and heat to approximately 50°C to dissolve completely.
2. Transfer 7.0 mL portions to test tubes and cap loosely.
3. Sterilize in the autoclave at 113–115°C for 20 minutes.
4. Remove from the autoclave and allow to cool to room temperature.

Materials

Six litmus milk tubes
18 to 24 hour pure cultures of:

Bacillus megaterium
Escherichia coli
Lactobacillus acidophilus
Morganella morganii
Streptococcus faecium

Test Protocol

1. Inoculate five tubes with the test cultures and leave one tube uninoculated as a control.
2. Label the tubes with the organisms' names, your name and the date.
3. Incubate the tubes aerobically with the uninoculated control at 35°C for 7 to 14 days.
4. Examine the tubes for characteristic color changes, gas production, and clot formation.
5. Record your results in the table below.

Precautions

⚠ Avoid overheating this medium and remove it from the autoclave immediately to avoid caramelization.

⚠ Because there are many possible reactions with this test, be sure to compare your results with an uninoculated control and with the table and photographs on pages 52–54 of the *Photographic Atlas.*

References

MacFaddin, Jean F. 1980. Page 194 in *Biochemical Tests for Identification of Medical Bacteria, 2nd Ed.* Williams & Wilkins, Baltimore, MD.

Power, David A. and Peggy J. McCuen. 1988. Page 177 in *Manual of BBL® Products and Laboratory Procedures, 6th Ed.* Becton Dickinson Microbiology Systems, Cockeysville, MD.

ORGANISM	RESULT
Bacillus megaterium	
Escherichia coli	
Lactobacillus acidophilus	
Morganella morganii	
Streptococcus faecium	

A = acid production; Alk = alkaline reaction; G = gas production; C = clot formation; R = reduction of litmus; P = proteolysis

Methyl Red and Voges-Proskauer (MRVP) Broth

Photographic Atlas Reference

Page 55

Recipes

MRVP Broth

Buffered peptone	7.0 g
Dipotassium phosphate	5.0 g
Dextrose (glucose)	5.0 g
Distilled or deionized water	1.0 L

final pH = 6.9 ± 0.2 at 25°C

Test Reagents

Methyl Red*

Methyl red dye	0.1 g
Ethanol	300.0 mL
Distilled water	200.0 mL

Barritt's Reagent A

α-naphthol	5.0 g
Absolute Ethanol	<100.0 mL**

Barritt's Reagent B

Potassium hydroxide	40.0 g
Distilled water	<100.0 mL**

*Mix dye in ethanol first, then add water to total 500.0 mL
**Total volume of solution is 100.0 mL

Materials

Three MRVP broths
Methyl red
Barritt's reagents A and B
Six nonsterile test tubes
Nonsterile 1 mL pipettes
18 to 24 hour pure cultures of:
Citrobacter diversus
Serratia marcescens

Medium Preparation

1. Suspend the ingredients in one liter of deionized or distilled water, mix well and warm until completely dissolved.
2. Transfer 7.0 mL portions to test tubes and cap loosely.
3. Sterilize in the autoclave at 15 lbs. pressure (121°C) for 15 minutes.
4. Remove from the autoclave and allow to cool to room temperature.

Test Protocol (See Figure 5-3)

1. Inoculate two broths with the test cultures and leave one uninoculated as a control.
2. Label the broths with the organisms' names, your name and the date.
3. Incubate the tubes aerobically with the uninoculated control at 35°C for 5 days.
4. After incubation, aseptically remove two 1.0 mL aliquots from each broth and place into separate tubes.
5. For each pair of tubes, add the reagents as follows:

   Tube #1
   a. Add several drops of methyl red reagent.
   b. Observe for red color formation.

   Tube #2 (See Precautions below.)
   a. Add 0.6 mL of Barritt's reagent A. Mix well.
   b. Add 0.2 mL of Barritt's reagent B. Mix well.
   c. Observe for red color formation within 5 minutes.
6. Record your results in the table below.

ORGANISM	MR +/-	VP +/-
Citrobacter diversus		
Serratia marcescens		

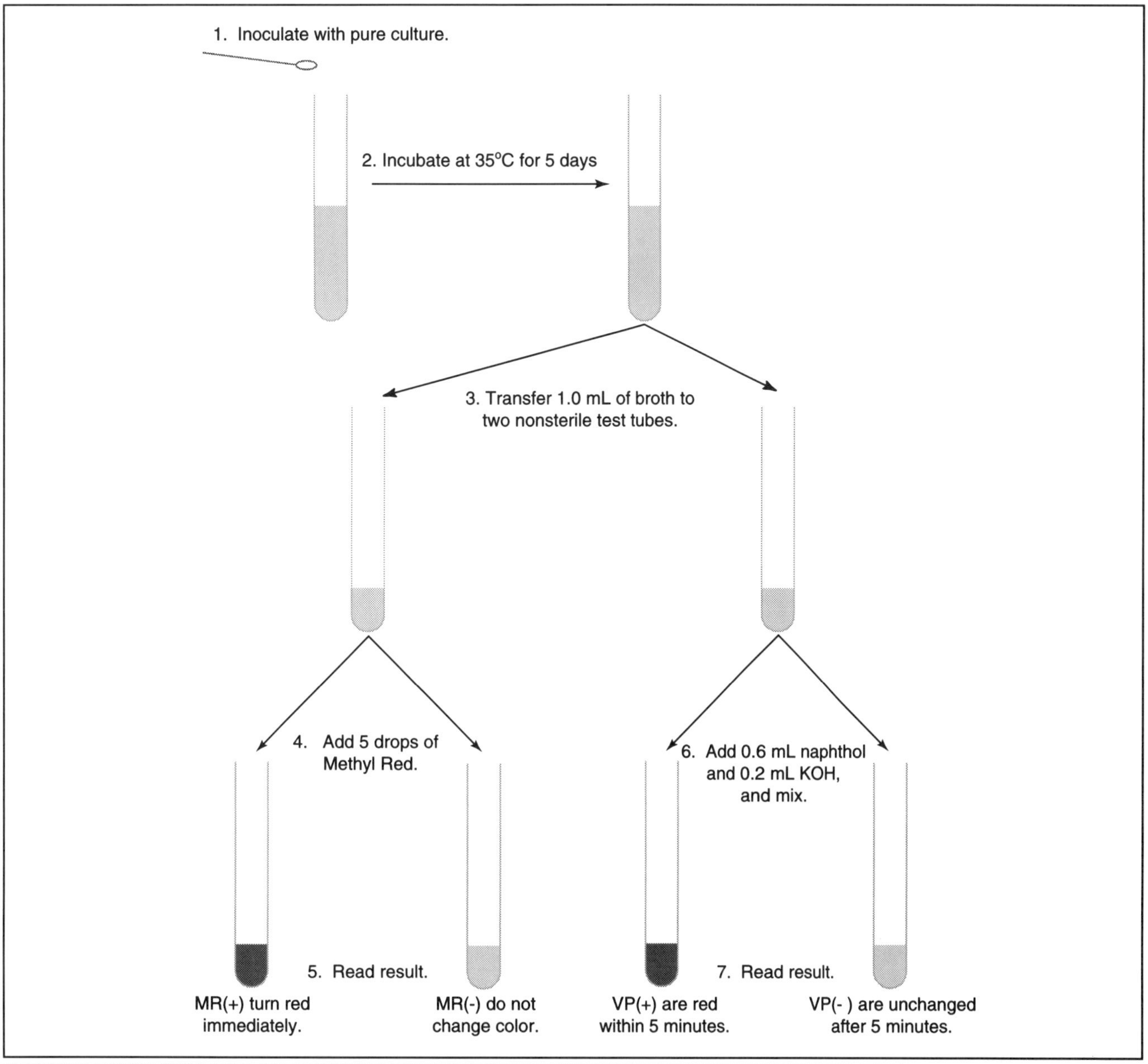

FIGURE 5-3.

Procedural Diagram *for the Methyl Red and Voges-Proskauer Tests.*

Precautions

⚠ It is essential that the Barritt's reagents be added in the correct order and that none of the measurements be exceeded. False Voges-Proskauer (VP) positives will likely occur if the KOH is added first.

⚠ A common problem with the VP test is differentiating weak positive reactions from negative reactions. VP (-) reactions often produce a copper color; weak VP (+) reactions produce a pink color. The strongest color change should be at the surface.

References

DIFCO Laboratories. 1984. Page 543 in *DIFCO Manual, 10th Ed.* DIFCO Laboratories, Detroit, MI.

MacFaddin, Jean F. 1980. Pages 209 and 308 in *Biochemical Tests for Identification of Medical Bacteria, 2nd Ed.* Williams & Wilkins, Baltimore, MD.

Power, David A. and Peggy J. McCuen. 1988. Page 202 in *Manual of BBL® Products and Laboratory Procedures, 6th Ed.* Becton Dickinson Microbiology Systems, Cockeysville, MD.

Smibert, Robert M. and Noel R. Krieg. 1994. Pages 622 and 630 in *Methods for General and Molecular Bacteriology*, edited by Philipp Gerhardt, R. G. E. Murray, Willis A. Wood and Noel R. Krieg, American Society for Microbiology, Washington, DC.

Milk Agar

Photographic Atlas Reference

Page 57

Recipe

Milk Agar

Beef extract	3.0 g
Peptone	5.0 g
Agar	15.0 g
Powdered nonfat milk	100.0 g
Distilled or deionized water	1.0 L

final pH = 7.2 ± 0.2 at 25°C

Preparation

1. Suspend the powdered milk in 500.0 mL of distilled or deionized water in a one liter flask, mix well and cover loosely.
2. Suspend the remainder of the ingredients in 500.0 mL of deionized water in a one liter flask, mix well, boil to dissolve completely and cover loosely.
3. Sterilize in the autoclave at 113–115°C for 20 minutes. (See Precautions below.)
4. Remove from the autoclave, allow to cool slightly, then aseptically pour the milk solution into the agar solution and mix *gently* (to prevent foaming).
5. Aseptically pour into sterile Petri dishes (15 mL/plate).
6. Allow to cool to room temperature.

Materials

One milk agar plate
18 to 24 hour pure cultures of:
Bacillus subtilis
Escherichia coli

Test Protocol

1. Using a marking pen, divide the plate into three equal sectors. Be sure to mark on the bottom of the plate.
2. Spot inoculate two sectors with the test organisms and leave the third sector uninoculated as a control.
3. Label the plate with the organisms' names, your name and the date.
4. Invert the plate and incubate it aerobically at 35°C for 24 hours.
5. Examine the plates for clearing around the bacterial growth.
6. Record your results in the table below.

Precaution

⚠ Avoid overheating this medium and remove it from the autoclave immediately to avoid caramelization.

References

Chan, E. C. S., Michael J. Pelczar, Jr. and Noel R Krieg. 1986. Page 137 in *Laboratory Exercises In Microbiology*, 5th Ed. McGraw-Hill Book Company.

DIFCO Laboratories. 1984. Page 619 in *DIFCO Manual, 10th Ed.* DIFCO Laboratories, Detroit, MI.

Holt, John G. (Editor). 1994. *Bergey's Manual of Determinative Bacteriology, 9th Ed.* Williams and Wilkins, Baltimore, MD.

Smibert, Robert M. and Noel R. Krieg. 1994. Page 613 in *Methods for General and Molecular Bacteriology*, edited by Philipp Gerhardt, R. G. E. Murray, Willis A. Wood and Noel R. Krieg, American Society for Microbiology, Washington, DC.

ORGANISM	RESULT +/–
Bacillus subtilis	
Escherichia coli	

Motility Agar

Photographic Atlas Reference

Page 58

Recipe

Motility Agar

Beef extract	3.0 g
Pancreatic digest of gelatin	10.0 g
Sodium chloride	5.0 g
Agar	4.0 g
Triphenyltetrazolium chloride (TTC)	0.05 g
Distilled or deionized water	1.0 L

final pH = 7.3 ± 0.2 at 25°C

Medium Preparation

1. Suspend the ingredients in one liter of distilled or deionized water, mix well and boil to dissolve completely.
2. Dispense 7.0 mL portions into test tubes and cap loosely.
3. Sterilize in the autoclave at 15 lbs. pressure (121°C) for 15 minutes.
4. Remove from the autoclave and allow to cool in the upright position.

Test Protocol

1. Stab inoculate two tubes with the test organisms. Stab the third tube with a sterile needle as a control.
2. Label the tubes with the organisms' names, your name and the date.
3. Incubate the tubes aerobically with the uninoculated control at 35°C for 24 to 48 hours.
4. Examine the growth pattern for characteristic spreading from the stab line. (See Figure 5-4.)

Materials

Three motility agar stabs
18 to 24 hour pure cultures of:
Enterobacter aerogenes
Staphylococcus aureus

Precaution

⚠ The bacterial growth pattern can be obscured by careless stabbing technique. Therefore, *gently* pull the inoculating needle out of the agar exactly the way it went in.

References

DIFCO Laboratories. 1984. Page 581 in *DIFCO Manual, 10th Ed.* DIFCO Laboratories, Detroit, MI.

MacFaddin, Jean F. 1980. Page 214 in *Biochemical Tests for Identification of Medical Bacteria, 2nd Ed.* Williams & Wilkins, Baltimore, MD.

Power, David A. and Peggy J. McCuen. 1988. Page 201 in *Manual of BBL® Products and Laboratory Procedures, 6th Ed.* Becton Dickinson Microbiology Systems, Cockeysville, MD.

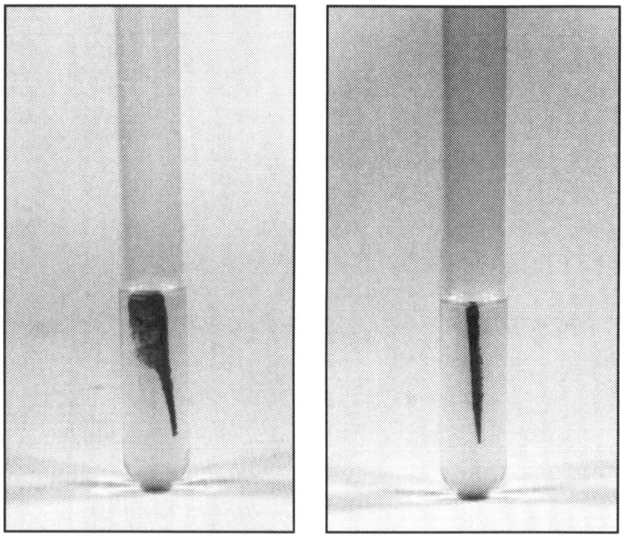

FIGURE 5-4.

Motile or Nonmotile? *These photographs, taken of the same motility agar stab, demonstrate one possible result of poor inoculation technique. Because of its spreading appearance, the growth pattern shown on the left might be interpreted as motility. The growth pattern on the right, seen after rotating the tube 90°, demonstrates that the organism is nonmotile. In this example, the student moved the needle to the side as she removed it from the agar. True motility is demonstrated by growth radiating in all directions from a single stab line.*

ORGANISM	RESULT +/–
Enterobacter aerogenes	
Staphylococcus aureus	

Nitrate Reduction Broth

Photographic Atlas Reference

Page 59

Recipes

Tryptic Nitrate Medium

Tryptose	20.0 g
Dextrose	1.0 g
Disodium phosphate	2.0 g
Potassium nitrate	1.0 g
Agar	1.0 g
Distilled or deionized water	1.0 L

final pH = 7.2 ± 0.2 at 25°C

Test Reagents

Reagent A

Sulfanilic acid	1.0 g
Acetic acid, 5N	125.0 mL

Reagent B

Dimethyl-α-naphthylamine	1.0 g
Acetic acid, 5N	200.0 mL

Materials

Five nitrate reduction broths
Test reagents A and B
Zinc powder
18 to 24 hour pure cultures of:
Acinetobacter calcoaceticus
Escherichia coli
Proteus mirabilis
Staphylococcus aureus

Medium Preparation

1. Suspend the ingredients in one liter of deionized or distilled water, mix well and boil until completely dissolved.
2. Transfer 10.0 mL portions to test tubes and cap loosely.
3. Sterilize in the autoclave at 15 lbs. pressure (121°C) for 15 minutes.
4. Remove from the autoclave and allow to cool to room temperature.

Test Protocol (See Figure 5-5)

1. Inoculate four broths with the test organisms and leave one uninoculated as a control.
2. Label the tubes with the organisms' names, your name and the date.
3. Incubate the tubes aerobically with the uninoculated control at 35°C for 24 to 48 hours.
4. After incubation, add the reagents as follows:
   a. Add 0.5 mL of Reagent A and 0.5 mL of reagent B to each tube. Mix well.
   b. Formation of a red color within approximately five minutes indicates reduction of nitrate to nitrite. Score as a (+1).
   c. If there is no color change, add a pinch of zinc dust. Formation of a red color within approximately ten minutes indicates the presence of nitrate, so the test is scored as negative for nitrate reduction.
   d. No color change after zinc addition indicates reduction of nitrate beyond nitrite, and is considered a positive test. Score as a (+2).
5. Record your results in the table below.

ORGANISM	RESULT
Acinetobacter calcoaceticus	
Escherichia coli	
Proteus mirabilis	
Staphylococcus aureus	

(-) = no reduction of nitrate; (+1) = reduction to nitrite; (+2) = reduction beyond nitrite

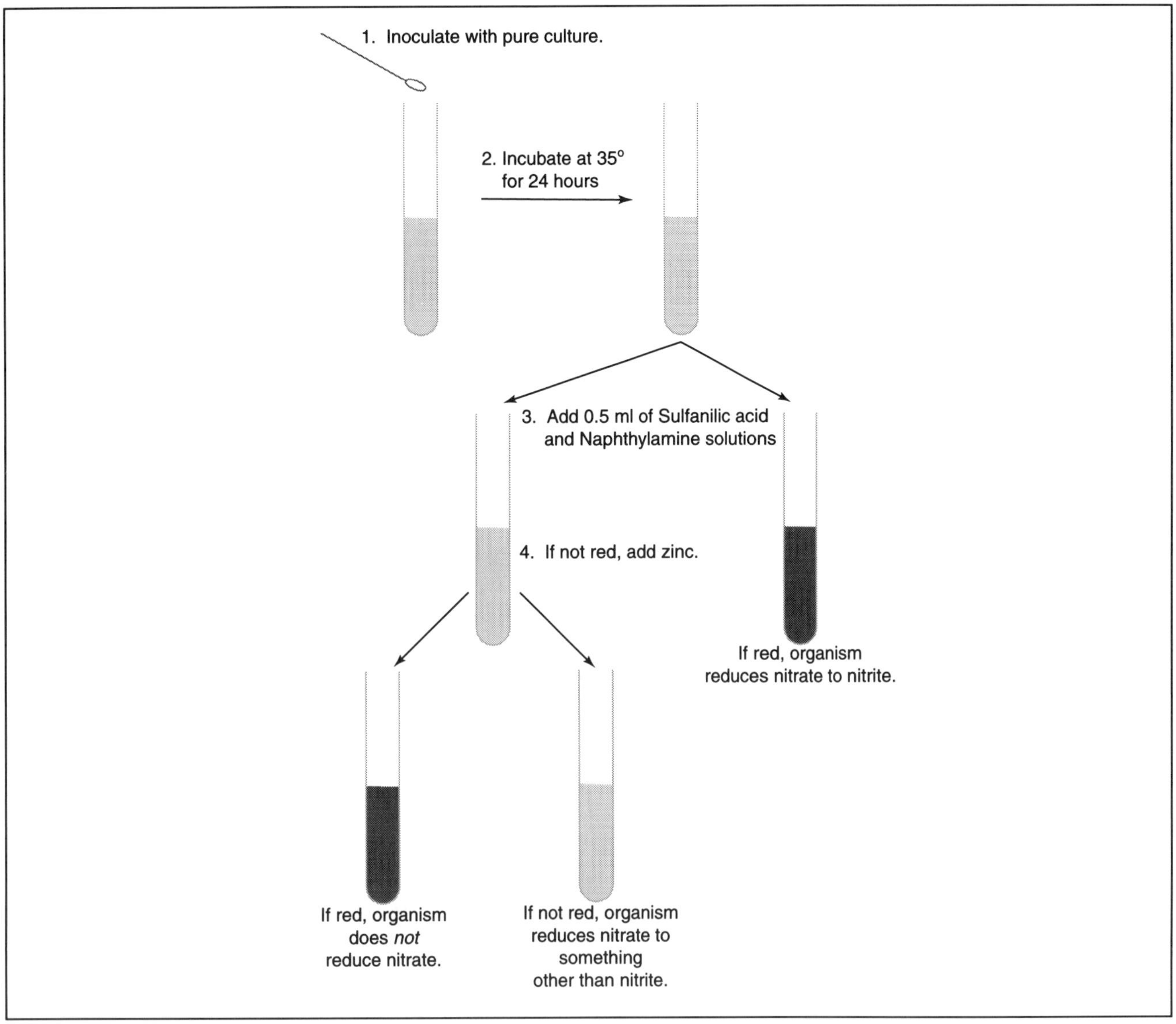

FIGURE 5-5.

Procedural Diagram *for the Nitrate Reduction Test.*

Precaution

⚠ The most common error is in misinterpreting the results. What does red mean *this* time?

References

DIFCO Laboratories. 1984. Page 1023 in *DIFCO Manual, 10th Ed.* DIFCO Laboratories, Detroit, MI.

Lányi, B. 1987. Page 21 in *Methods in Microbiology, Vol. 19*, edited by R. R. Colwell and R. Grigorova, Academic Press Inc., New York.

MacFaddin, Jean F. 1980. Page 236 in *Biochemical Tests for Identification of Medical Bacteria, 2nd Ed.* Williams & Wilkins, Baltimore, MD.

Power, David A. and Peggy J. McCuen. 1988. Page 213 in *Manual of BBL® Products and Laboratory Procedures, 6th Ed.* Becton Dickinson Microbiology Systems, Cockeysville, MD.

o-Nitrophenyl-ß-D-Galactopyranoside (ONPG) Test

Photographic Atlas Reference

Page 61

Test Protocol

The following procedure is the one recommended by Key Scientific for their product. If you use a different medium, the procedure may differ slightly. Be sure to read the instructions carefully before performing the test.

Materials

Key Scientific Products' K490 ONPG Tablets
Three small sterile test tubes (12 mm x 75 mm)
Sterile distilled or deionized water
Sterile 1.0 mL pipettes
18 to 24 hour pure cultures on solid medium:
- *Enterobacter aerogenes*
- *Salmonella typhimurium*

1. Dispense 1.0 mL volumes of sterile distilled or deionized water into three small sterile test tubes.
2. Add one test tablet to each tube and mix well to dissolve it completely.
3. Inoculate two ONPG tubes, leaving the third tube as a control.
4. Label the tubes with the organisms' names, your name and the date.
5. Incubate the tubes aerobically with the uninoculated control at 35°C for up to 6 hours, or until the medium turns yellow.
6. Record your results in the table below.

Precaution

⚠ Allow the tubes to incubate the full 6 hours before scoring them as negative.

Reference

Package insert from Key Scientific Products' K490 ONPG Tablets.

ONPG test procedure courtesy of Key Scientific Products.

ORGANISM	RESULT +/–
Enterobacter aerogenes	
Salmonella typhimurium	

Oxidation-Fermentation (OF) Medium

Photographic Atlas Reference

Page 63

Recipes

OF Basal Medium

Pancreatic digest of casein	2.0 g
Sodium chloride	5.0 g
Dipotassium phosphate	0.3 g
Agar	2.5 g
Bromthymol blue	0.03 g
Distilled or deionized water	1.0 L

final pH = 6.8 ± 0.1 at 25°C

Carbohydrate Solution

Carbohydrate (glucose, lactose, sucrose)	1.0 g
Distilled or deionized water	<10.0 mL*

*Total solution volume is 10.0 mL

Medium Preparation

1. Suspend the ingredients, *without the carbohydrate*, in one liter of distilled or deionized water, mix well and boil to dissolve completely. This is basal medium.
2. Divide the medium into ten aliquots of 100.0 mL each.
3. Cover loosely and sterilize in the autoclave at 121°C for 15 minutes.
4. Cool to 50°C.
5. Prepare 10% solutions of carbohydrates to be tested.
6. Sterilize in the autoclave at 118°C for 10 minutes.
7. Aseptically add 10.0 mL sterile carbohydrate solution to a 50°C basal medium aliquot and mix well.
8. Aseptically transfer 7.0 mL volumes to sterile test tubes and allow to cool.

Materials

Ten OF tubes
Sterile mineral oil
Sterile Pasteur pipettes
18 to 24 hour pure cultures of:
Alcaligenes faecalis
Acinetobacter calcoaceticus
Enterobacter aerogenes
Pseudomonas aeruginosa

Test Protocol

1. Heat all the OF tubes in boiling water for approximately ten minutes to remove the dissolved oxygen.
2. Cool the tubes quickly to room temperature by placing them in cold tap water.
3. Stab inoculate 2 OF tubes for each organism being tested. Using a light inoculum, stab to a depth of about 0.5 cm from the bottom of the agar. Leave two tubes uninoculated as controls.
4. Overlay one of each pair of tubes with 2-3 mL of sterile mineral oil. Overlay one of the control tubes as well. (See Figure 5-6.)
5. Label the tubes with the organisms' names, your name and the date.
6. Incubate the tubes aerobically with both uninoculated controls at 35°C for 48 hours.
7. Examine the tubes for characteristic yellow color formation. Reincubate any OF tubes that have not changed color for an additional 48 hours.
8. Record your results in the table on the next page.

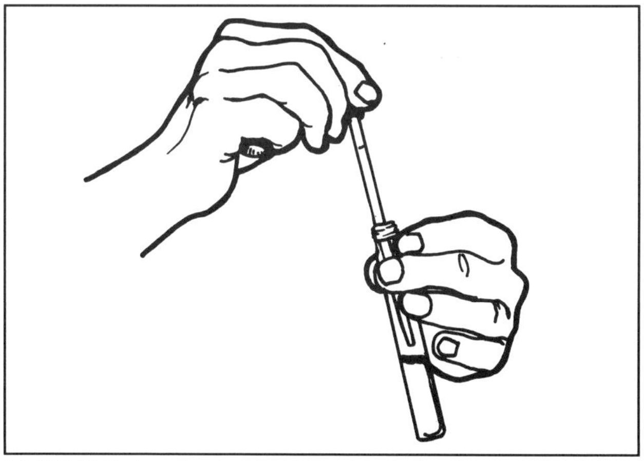

FIGURE 5-6.

Adding the Mineral Oil Layer *Tip the tube slightly to one side and gently add 2 to 3 mL of sterile mineral oil. Be sure to use a sterile pipette for each tube.*

Precaution

⚠ Slight yellowing may occur at the top of some overlaid tubes due to the diffusion of oxygen through the mineral oil layer. This should not be confused with a fermentation reaction. Refer to the *Photographic Atlas* to see the comparative differences.

References

Collins, C. H., Patricia M. Lyne, J. M. Grange. 1995. Page 112 in *Collins and Lyne's Microbiological Methods, 7th Ed.* Butterworth-Heinemann, UK.

DIFCO Laboratories. 1984. Page 625 in *DIFCO Manual, 10th Ed.* DIFCO Laboratories, Detroit, MI.

MacFaddin, Jean F. 1980. Page 260 in *Biochemical Tests for Identification of Medical Bacteria, 2nd Ed.* Williams & Wilkins, Baltimore, MD.

Power, David A. and Peggy J. McCuen. 1988. Page 216 in *Manual of BBL® Products and Laboratory Procedures, 6th Ed.* Becton Dickinson Microbiology Systems, Cockeysville, MD.

Smibert, Robert M. and Noel R. Krieg. 1994. Page 625 in *Methods for General and Molecular Bacteriology*, edited by Philipp Gerhardt, R. G. E. Murray, Willis A. Wood and Noel R. Krieg, American Society for Microbiology, Washington, DC.

ORGANISM	GLUCOSE		SUCROSE		LACTOSE	
	SEALED	UNSEALED	SEALED	UNSEALED	SEALED	UNSEALED
Acinetobacter calcoaceticus						
Alcaligenes faecalis						
Enterobacter aerogenes						
Pseudomonas aeruginosa						

O = oxidative metabolism; F = fermentation; N = nonsaccharolytic (no sugar metabolism)

Oxidase Test

Photographic Atlas Reference

Page 64

Recipes

Nutrient Agar

Beef extract	3.0 g
Peptone	5.0 g
Agar	15.0 g
Distilled or deionized water	1.0 L
final pH = 6.8 ± 0.2 at 25°C	

Test Reagent

Tetramethyl-*p*-phenylenediamine dihydrochloride	1.0 g
Deionized Water	100.0 mL

Medium Preparation

1. Suspend, mix and boil the ingredients in one liter of distilled or deionized water to dissolve completely.
2. Cover loosely and sterilize in the autoclave at 15 lbs. pressure (121°C) for 15 minutes.
3. Aseptically pour into sterile Petri dishes (15 mL/plate) and allow to cool to room temperature.

Materials

One nutrient agar plate
Test reagent
Filter paper (7 cm diameter or smaller) in a Petri dish
Sterile toothpicks or cotton swabs
Pasteur pipettes
18 to 24 hour pure cultures of:
Escherichia coli
Moraxella catarrhalis

Test Protocol

Day one

1. Using a marking pen, divide the nutrient agar plate into three equal sectors. Be sure to mark the bottom of the plate.
2. Spot inoculate two sectors with the test organisms and leave the third sector as a control.
3. Label the plate with the organisms' names, your name and the date.
4. Invert the plate and incubate it aerobically at 35°C for 24 hours.

Day two

Indirect Method (See Precautions below.)

1. Place the filter paper in the sterile Petri dish and saturate it with the test reagent.
2. Using a sterile toothpick or cotton swab transfer a large amount of fresh growth from the plate to the filter paper.
3. Observe for color change within 10 to 15 seconds.
4. Record your results in the table below.

Direct Method (See Precautions below.)

1. Place a few drops of oxidase test reagent onto the bacterial growth and onto the uninoculated control sector.
2. Observe for color change within 10 to 15 seconds.
3. Record your results in the table below.

Precautions

⚠ Tetramethyl-*p*-phenylenediamine in solution is very unstable and will auto-oxidize after a short time. Therefore, color changes after 60 seconds are not considered positive.

ORGANISM	DIRECT +/-	INDIRECT +/-
Escherichia coli		
Moraxella catarrhalis		

⚠ The test reagent should be made immediately prior to its use. If it must be stored, keep it in the refrigerator or freezer and allow it to warm to room temperature immediately before use.

References

Collins, C. H., Patricia M. Lyne, J. M. Grange. 1995. Page 116 in *Collins and Lyne's Microbiological Methods, 7th Ed.* Butterworth-Heinemann, UK.

Lányi, B. 1987. Page 18 in *Methods in Microbiology, Vol. 19*, edited by R. R. Colwell and R. Grigorova, Academic Press Inc., New York.

MacFaddin, Jean F. 1980. Page 249 in *Biochemical Tests for Identification of Medical Bacteria, 2nd Ed.* Williams & Wilkins, Baltimore, MD.

Smibert, Robert M. and Noel R. Krieg. 1994. Page 625 in *Methods for General and Molecular Bacteriology*, edited by Philipp Gerhardt, R. G. E. Murray, Willis A. Wood and Noel R. Krieg, American Society for Microbiology, Washington, DC.

Phenol Red Fermentation Broth

Photographic Atlas Reference

Page 66

Recipe

Phenol Red (Carbohydrate) Broth

Pancreatic digest of casein	10.0 g
Sodium chloride	5.0 g
Carbohydrate (glucose, lactose, or sucrose)	5.0 g
Phenol red	0.018 g
Distilled or deionized water	1.0 L

final pH = 7.3 ± 0.2 at 25°C

Medium Preparation

1. Suspend the ingredients in one liter of distilled or deionized water, mix well and warm slightly to dissolve completely. (Use only one of the listed carbohydrates.)
2. Dispense 7.0 mL volumes into test tubes.
3. Insert inverted Durham tubes into the test tubes and cap loosely.
4. Sterilize in the autoclave at 116–118°C for 15 minutes. (See Precautions below.)
5. Remove from the autoclave and allow to cool to room temperature.

Materials

Four broths for each carbohydrate being tested
18 to 24 hour pure cultures of:
- *Klebsiella pneumoniae*
- *Pseudomonas aeruginosa*
- *Streptococcus faecalis*

Test Protocol

1. Inoculate three broths of each carbohydrate with the test organisms. Leave one of each carbohydrate broth uninoculated as a control. (See Precautions below.)
2. Label the tubes with the organisms' names, your name and the date.
3. Incubate all the tubes (including the uninoculated control of each carbohydrate broth) aerobically at 35°C for 24 to 48 hours.
4. Examine the tubes for acid and/or gas production or alkaline reactions.
5. Record your results in the table below.

Precautions

⚠ Avoid overheating this medium and remove it from the autoclave immediately after sterilization.

⚠ Don't forget to compare your results with the uninoculated control. This test is sometimes difficult to read, especially with weak alkaline reactions. The best results are obtained when inoculating with fresh bacterial cultures.

⚠ Some actively fermenting bacteria will cause a reversion from acidity to alkalinity of this medium after 48 hours. Therefore, increase the carbohydrate to 1% (10 g/L) if you plan to incubate for a longer time.

References

DIFCO Laboratories. 1984. Page 660 in *DIFCO Manual, 10th Ed.* DIFCO Laboratories, Detroit, MI.

Lányi, B. 1987. Page 44 in *Methods in Microbiology, Vol. 19*, edited by R. R. Colwell and R. Grigorova, Academic Press Inc., New York.

MacFaddin, Jean F. 1980. Page 36 in *Biochemical Tests for Identification of Medical Bacteria, 2nd Ed.* Williams & Wilkins, Baltimore, MD.

Power, David A. and Peggy J. McCuen. 1988. Page 220 in *Manual of BBL® Products and Laboratory Procedures, 6th Ed.* Becton Dickinson Microbiology Systems, Cockeysville, MD.

ORGANISM	PR GLUCOSE	PR LACTOSE	PR SUCROSE
Klebsiella pneumoniae			
Pseudomonas aeruginosa			
Streptococcus faecalis			

A = acid production; Alk = alkaline reaction; G = gas production

Phenylalanine Deaminase Agar

Photographic Atlas Reference

Page 69

Recipes

Phenylalanine Deaminase Agar

DL-Phenylalanine	2.0 g
Yeast extract	3.0 g
Sodium chloride	5.0 g
Sodium phosphate	1.0 g
Agar	12.0 g
Distilled or deionized water	1.0 L

final pH =7.3 ± 0.2 at 25°C

Test Reagent

Ferric chloride	10.0 g
Deionized water	<100.0 mL *

*Total solution volume is 100.0 mL

Medium Preparation

1. Suspend the ingredients in one liter of distilled or deionized water, mix well and boil to dissolve completely.
2. Dispense 7.0 mL volumes into test tubes and cap loosely.
3. Sterilize in the autoclave at 15 lbs pressure (121°C) for 10 minutes.
4. Remove from the autoclave, slant and allow to cool to room temperature.

Materials

Three phenylalanine deaminase agar slants
Test reagent
18 to 24 hour pure cultures of:
Enterobacter aerogenes
Proteus vulgaris

Test Protocol

1. Streak two slants with the test organisms leaving one slant uninoculated as a control.
2. Label the tubes with the organisms' names, your name and the date.
3. Incubate the slants aerobically with the uninoculated control at 35°C for 18 to 24 hours.
4. Add a few drops of test reagent to each tube.
5. Observe for the characteristic green color formation within 1 to 5 minutes.
6. Record your results in the table below.

Precaution

⚠ The green color resulting from positive reactions in this test will fade quickly so read and record your results immediately.

References

DIFCO Laboratories. 1984. Page 664 in *DIFCO Manual, 10th Ed.* DIFCO Laboratories, Detroit, MI.

Lányi, B. 1987. Page 28 in *Methods in Microbiology, Vol. 19*, edited by R. R. Colwell and R. Grigorova, Academic Press Inc., New York.

MacFaddin, Jean F. 1980. Page 269 in *Biochemical Tests for Identification of Medical Bacteria, 2nd Ed.* Williams & Wilkins, Baltimore, MD.

Power, David A. and Peggy J. McCuen. 1988. Page 222 in *Manual of BBL® Products and Laboratory Procedures, 6th Ed.* Becton Dickinson Microbiology Systems, Cockeysville, MD.

ORGANISM	RESULT +/-
Enterobacter aerogenes	
Proteus vulgaris	

Sulfur-Indole-Motility (SIM) Medium

Photographic Atlas Reference

Page 70

Recipes

SIM Medium

Pancreatic digest of casein	20.0 g
Peptic digest of animal tissue	6.1 g
Ferrous ammonium sulfate	0.2 g
Sodium thiosulfate	0.2 g
Agar	3.5 g
Distilled or deionized water	1.0 L

final pH =7.3 ± 0.2 at 25°C

Test Reagent (Kovac's Reagent)

Amyl alcohol	75.0 mL
Hydrochloric acid, concentrated	25.0 mL
p-Dimethylaminobenzaldehyde	5.0 g

Medium Preparation

1. Suspend the ingredients in one liter of distilled or deionized water, mix well and boil to dissolve completely.
2. Dispense 7.0 mL volumes into test tubes and cap loosely.
3. Sterilize in the autoclave at 15 lbs pressure (121°C) for 15 minutes.
4. Remove from the autoclave and allow to cool to room temperature

Materials

Four SIM tubes
Kovac's reagent
18 to 24 hour pure cultures of:
Escherichia coli
Salmonella typhimurium
Shigella flexneri

Test Protocol

1. Stab inoculate three SIM tubes with the test organisms leaving the fourth uninoculated as a control.
2. Label the tubes with the organisms' names, your name and the date.
3. Incubate the tubes aerobically with the uninoculated control at 35°C for 24 to 48 hours.
4. Examine the tubes for formation of black precipitate in the medium *and* spreading from the stab line. Record any H_2S production and/or motility in the table below.
5. Add a few drops of Kovac's reagent to each tube.
6. Observe for the formation of a red color in the reagent layer.
7. Record your results in the table below.

Precaution

⚠ Kovac's reagent should be stored in the refrigerator. Although commercial brands include an expiration date, note the reagent's color. A color change from pale yellow to brown is an indication that the solution has deteriorated.

References

DIFCO Laboratories. 1984. Page 762 in *DIFCO Manual, 10th Ed.* DIFCO Laboratories, Detroit, MI.

MacFaddin, Jean F. 1980. Page 162 in *Biochemical Tests for Identification of Medical Bacteria, 2nd Ed.* Williams & Wilkins, Baltimore, MD.

Power, David A. and Peggy J. McCuen. 1988. Page 246 in *Manual of BBL® Products and Laboratory Procedures, 6th Ed.* Becton Dickinson Microbiology Systems, Cockeysville, MD.

ORGANISM	H_2S +/–	INDOLE +/–	MOTILITY +/–
Escherichia coli			
Salmonella typhimurium			
Shigella flexneri			

Starch Agar

Photographic Atlas Reference

Page 73

Recipe

Starch Agar

Beef extract	3.0 g
Soluble starch	10.0 g
Agar	12.0 g
Distilled or deionized water	1.0 L

final pH = 7.5 ± 0.2 at 25°C

Test Reagent (Gram Iodine)

Use the iodine in your Gram stain kit.

Medium Preparation

1. Suspend the ingredients in one liter of distilled or deionized water, mix well and boil to dissolve completely.
2. Sterilize in the autoclave at 15 lbs pressure (121°C) for 15 minutes.
3. Remove from the autoclave and allow to cool slightly.
4. Aseptically pour into sterile Petri dishes (15 mL per plate).

Materials

One starch agar plate
Gram iodine
18 to 24 hour pure cultures of:
Bacillus subtilis
Staphylococcus aureus

Test Protocol

1. Using a marking pen, divide the starch agar plate into three equal sectors. Be sure to mark on the bottom of the plate.
2. Spot inoculate two sectors with the test organisms leaving the third sector as a control.
3. Label the plate with the organisms' names, your name and the date.
4. Invert the plate and incubate it aerobically at 35°C for 48 hours.
5. Flood the plate with Gram iodine. (See Precautions below.)
6. Examine the plate for clearing around the growth.
7. Record your results in the table below.

Precautions

⚠ Gram Iodine is poisonous and should be treated with care.

⚠ Note the location and appearance of the growth *before* adding the iodine. Occasionally, growth that is thinning at the edge will give the appearance of clearing.

References

Collins, C. H., Patricia M. Lyne, J. M. Grange. 1995. Page 117 in *Collins and Lyne's Microbiological Methods, 7th Ed.* Butterworth-Heinemann, UK.

DIFCO Laboratories. 1984. Page 879 in *DIFCO Manual, 10th Ed.*, DIFCO Laboratories, Detroit, MI.

Lányi, B. 1987. Page 55 in *Methods in Microbiology, vol. 19*, edited by R. R. Colwell and R. Grigorova, Academic Press Inc., New York.

MacFaddin, Jean F. 1980. Page 286 in *Biochemical Tests for Identification of Medical Bacteria, 2nd Ed.* Williams & Wilkins, Baltimore, MD.

Smibert, Robert M. and Noel R. Krieg. 1994. Page 630 in *Methods for General and Molecular Bacteriology*, edited by Philipp Gerhardt, R. G. E. Murray, Willis A. Wood and Noel R. Krieg, American Society for Microbiology, Washington, DC.

ORGANISM	RESULT +/-
Bacillus subtilis	
Staphylococcus aureus	

Tribubtyrin Agar

Photographic Atlas Reference

Page 75

Recipe

Tributyrin Agar

Beef extract	1.5 g
Peptone	2.5 g
Agar	7.5 g
Tributyrin oil	5.0 mL
Distilled or deionized water	500.0 mL

final pH = 6.0 ± 0.2 at 25°C

Medium Preparation

1. Suspend the dry ingredients in 500.0 mL of deionized water, mix well and boil to dissolve completely.
2. Cover loosely and sterilize together with the tube of tributyrin oil in the autoclave at 15 lbs pressure (121°C) for 15 minutes.
3. Remove from the autoclave and aseptically pour the agar mixture into a sterile glass blender.
4. Aseptically add the tributyrin oil to the agar mixture and blend on "High" for 1 minute.
5. Aseptically pour into sterile Petri dishes (15 mL/plate).

Materials

One tributyrin agar plate
18 to 24 hour pure cultures of:
Enterobacter aerogenes
Moraxella catarrhalis

Test Protocol

1. Using a marking pen, divide the tributyrin agar plate into three equal sectors. Be sure to mark on the bottom of the plate.
2. Spot inoculate two sectors with the test organisms leaving the third sector uninoculated as a control.
3. Label the plate with the organisms' names, your name and the date.
4. Invert the plate and incubate it aerobically at 35°C for 48 hours.
5. Examine the plate for clearing around the bacterial growth.
6. Record your results in the table below.

Precaution

⚠ The tributyrin oil in the medium will etch the plastic petri dishes after a few days. Since this makes reading the plates more difficult, we recommend using these plates within 48 hours after preparation.

References

Collins, C. H., Patricia M. Lyne, J. M. Grange. 1995. Page 114 in *Collins and Lyne's Microbiological Methods, 7th Ed.* Butterworth-Heinemann, UK.

DIFCO Laboratories. 1984. Page 619 in *DIFCO Manual, 10th Ed.* DIFCO Laboratories, Detroit, MI.

Knapp, Joan S. and Roselyn J. Rice. 1995. Page 335 in *Manual of Cinical. Microbiology, 6th Ed.*, edited by Patrick R. Murray, Ellen Jo Baron, Michael A. Pfaller, Fred C. Tenover and Robert H. Yolken, ASM Press, Washington, DC.

ORGANISM	RESULT +/–
Enterobacter aerogenes	
Moraxella catarrhalis	

Triple Sugar Iron (TSI) Agar

Photographic Atlas Reference

Page 77

Recipe

Triple Sugar Iron Agar

Beef extract	3.0 g
Yeast extract	3.0 g
Peptone	15.0 g
Proteose peptone	5.0 g
Dextrose (glucose)	1.0 g
Lactose	10.0 g
Sucrose	10.0 g
Ferrous sulfate	0.2 g
Sodium chloride	5.0 g
Sodium thiosulfate	0.3 g
Agar	12.0 g
Phenol red	0.024 g
Distilled or deionized water	1.0 L

final pH = 7.4 ± 0.2 at 25°C

Medium Preparation

1. Suspend the ingredients in one liter of distilled or deionized water, mix well and boil to dissolve completely.
2. Transfer 7.0 mL portions to test tubes and cap loosely.
3. Sterilize in the autoclave at 121°C for 15 minutes.
4. Slant so that the butt portion of the tube remains approximately 3 cm thick.
5. Allow to cool to room temperature.

Materials

Five TSI slants
18 to 24 hour pure cultures on solid media:

Escherichia coli
Proteus vulgaris
Pseudomonas aeruginosa
Salmonella typhimurium

Test Protocol

1. Stab inoculate four TSI slants with the test organisms using a heavy inoculum. As you withdraw the needle, streak the entire slant. (See Precautions below.)
2. Label the tubes with the organisms' names, your name and the date.
3. Incubate the slants aerobically with the uninoculated control at 35°C for no less than 18 hours or more than 24 hours.
4. Examine the tubes for characteristic color changes and gas production.
5. Record your results in the table below.

Precautions

⚠ The inoculation of butt and slant can be done in either order but both methods require a heavy inoculum.

⚠ Time is critical in this test; it must be read between 18 and 24 hours after inoculation.

⚠ If gas is produced it will appear as bubbles or fissures in the medium or will force the entire agar butt off the bottom of the tube.

References

DIFCO Laboratories. 1984. Page 1019 in *DIFCO Manual, 10th Ed.* DIFCO Laboratories, Detroit, MI.

Lányi, B. 1987. Page 44 in *Methods in Microbiology, Vol. 19*, edited by R. R. Colwell and R. Grigorova, Academic Press Inc., New York.

MacFaddin, Jean F. 1980. Page 183 in *Biochemical Tests for Identification of Medical Bacteria, 2nd Ed.* Williams & Wilkins, Baltimore, MD.

Power, David A. and Peggy J. McCuen. 1988. Page 269 in *Manual of BBL® Products and Laboratory Procedures, 6th Ed.* Becton Dickinson Microbiology Systems, Cockeysville, MD.

ORGANISM	RESULT
Escherichia coli	
Proteus vulgaris	
Pseudomonas aeruginosa	
Salmonella typhimurium	

A = acid production; Alk = alkaline reaction; H_2S = H_2S production; G = gas production; (–) = no reaction

Urease Agar

Photographic Atlas Reference

Page 79

Recipe

Urease Agar

Peptone	1.0 g
Dextrose (glucose)	1.0 g
Sodium chloride	5.0 g
Potassium phosphate, monobasic	2.0 g
Agar	15.0 g
Phenol red	0.012 g
Distilled or deionized water	1.0 L

final pH = 6.8 ± 0.2 at 25°C

Medium Preparation

1. Suspend the agar in 900 mL distilled or deionized water, mix well and boil to dissolve completely.
2. Cover loosely and sterilize by autoclaving at 15 lbs. pressure (121°C) for 15 minutes.
3. Remove from the autoclave and allow to cool to 55°C.
4. Suspend the remaining ingredients in 100 mL distilled or deionized water, mix well and filter sterilize. **Do not autoclave.** This is urease agar base.
5. Aseptically add the urease agar base to the agar solution and mix well.
6. Aseptically transfer 7.0 mL portions to sterile test tubes and cap loosely.
7. Slant in such a way that the agar butt is approximately twice as long as the slant.
8. Allow to cool to room temperature.

Materials

Three urease agar slants
18 to 24 hour pure cultures on solid media:
Enterobacter aerogenes
Proteus vulgaris

Test Protocol

1. Streak inoculate two slants with the test organisms, covering the entire agar surface with a heavy inoculum. Do not stab the butt.
2. Label the tubes with the organisms' names, your name and the date.
3. Incubate aerobically with the uninoculated control at 35°C for 6 days.
4. Observe the slants after 2 hours, 6 hours and at 24 hour intervals thereafter.
5. Using the symbols provided, record your observations in the table below.

Precaution

⚠ This test not only demonstrates the presence or absence of urease, but allows comparison of urease activity rates. Therefore, it is important to inoculate all the slants at the same time and to read all the results at the same time.

References

Collins, C. H., Patricia M. Lyne, J. M. Grange. 1995. Page 117 in *Collins and Lyne's Microbiological Methods, 7th Ed.* Butterworth-Heinemann, UK.

DIFCO Laboratories. 1984. Page 1040 in *DIFCO Manual, 10th Ed.* DIFCO Laboratories, Detroit, MI.

Lányi, B. 1987. Page 24 in *Methods in Microbiology, Vol. 19*, edited by R. R. Colwell and R. Grigorova, Academic Press Inc., New York.

MacFaddin, Jean F. 1980. Page 298 in *Biochemical Tests for Identification of Medical Bacteria, 2nd Ed.* Williams & Wilkins, Baltimore, MD.

Power, David A. and Peggy J. McCuen. 1988. Page 280 in *Manual of BBL® Products and Laboratory Procedures, 6th Ed.* Becton Dickinson Microbiology Systems, Cockeysville, MD.

ORGANISM	2 HRS.	6 HRS.	24 HRS.	2 DAYS	3 DAYS	4 DAYS	5 DAYS	6 DAYS
Enterobacter aerogenes								
Proteus vulgaris								

(+4) = entire tube is pink; (+2) = only the slant is pink; (+) = tip of the slant is pink; (–) = no pink

Urease Broth

Photographic Atlas Reference

Page 79

Recipe

Urease Broth

Yeast extract	0.1 g
Potassium phosphate, monobasic	9.1 g
Potassium phosphate, dibasic	9.5 g
Urea	20.0 g
Phenol red	0.01 g
Distilled or deionized water	1.0 L

final pH = 6.8 ± 0.2 at 25°C

Medium Preparation

1. Suspend the ingredients in one liter distilled or deionized water and mix well.
2. Filter sterilize the solution. **Do not autoclave.**
3. Aseptically transfer 1.0 mL volumes to small sterile test tubes and cap loosely.

Materials

Three urease broths
18 to 24 hour pure cultures on solid media:

Enterobacter aerogenes
Proteus vulgaris

Test Protocol

1. Inoculate two broths with a heavy inoculum from the test organisms. Leave the third broth uninoculated as a control.
2. Label the tubes with the organisms' names, your name and the date.
3. Incubate the tubes with the uninoculated control at 35°C for 24 to 48 hours.
4. Observe the tubes for color changes.
5. Record your results in the table below.

Precaution

⚠ Do not heat this broth above 50°C.

References

Collins, C. H., Patricia M. Lyne, J. M. Grange. 1995. Page 117 in *Collins and Lyne's Microbiological Methods, 7th Ed.* Butterworth-Heinemann, UK.

DIFCO Laboratories. 1984. Page 1043 in *DIFCO Manual, 10th Ed.* DIFCO Laboratories, Detroit, MI.

MacFaddin, Jean F. 1980. Page 298 in *Biochemical Tests for Identification of Medical Bacteria, 2nd Ed.* Williams & Wilkins, Baltimore, MD.

Power, David A. and Peggy J. McCuen. 1988. Page 281 in *Manual of BBL® Products and Laboratory Procedures, 6th Ed.* Becton Dickinson Microbiology Systems, Cockeysville, MD.

Smibert, Robert M. and Noel R. Krieg. 1994. Page 630 in *Methods for General and Molecular Bacteriology*, edited by Philipp Gerhardt, R. G. E. Murray, Willis A. Wood and Noel R. Krieg, American Society for Microbiology, Washington, DC.

ORGANISM	RESULT +/-
Enterobacter aerogenes	
Proteus vulgaris	

SECTION SIX

Quantitative Techniques

Viable Count

Photographic Atlas Reference

Page 81

Recipes

Nutrient Agar

Beef extract	3.0 g
Peptone	5.0 g
Agar	15.0 g
Distilled or deionized water	1.0 L
final pH = 6.8 ± 0.2 at 25°C	

Dilution Tubes

NaCl	0.9 g
Distilled or deionized water	<100.0 mL*

*Total solution volume is 100.0 mL.

Materials

Sterile 1 mL pipettes and pipettor
Two 9.9 mL dilution tubes
Three 9.0 mL dilution tubes
Eight nutrient agar plates
Beaker containing ethanol and a bent glass rod
Hand tally counter
Colony counter
24 hour broth culture of *Escherichia coli*

Medium Preparation

Nutrient Agar

1. Suspend the ingredients in one liter of distilled or deionized water, mix well and boil to dissolve completely.
2. Cover loosely and sterilize in the autoclave at 15 lbs. pressure (121°C) for 15 minutes.
3. Remove from the autoclave, allow to cool slightly and aseptically pour into sterile Petri dishes (15 mL/plate).
4. Allow to cool to room temperature.

Dilution Tubes

1. Dissolve the NaCl in approximately 90 mL of water.
2. Add water to bring the total volume up to 100.0 mL.
3. Cover loosely and sterilize in the autoclave at 121°C for 15 minutes.
4. Allow to cool to room temperature.
5. Aseptically dispense into sterile test tubes in the amounts of 9.0 mL and 9.9 mL.

Test Protocol (See Figure 6-1a.)

The following procedure includes inoculation by the "spread plate techinque." For instruction on this technique refer to page 115 in Appendix B.

1. Organize the plates into 4 pairs and label them A, B, C and D.
2. Aseptically transfer 0.1 mL from the broth culture (the *original sample*) to a 9.9 mL dilution tube and mix well. This is *DF* (dilution factor) *10^{-2}*. (For a discussion of dilution factors, see Precautions below and pages 81 and 82 of the *Photographic Atlas.*)
3. Aseptically transfer 0.1 mL DF 10^{-2} to a 9.9 mL dilution tube. Mix well. This is *DF 10^{-4}*.
4. Aseptically transfer 1.0 mL DF 10^{-4} to a 9.0 mL dilution tube. Mix well. This is *DF 10^{-5}*.
5. Aseptically transfer 1.0 mL DF 10^{-5} to a 9.0 mL dilution tube. Mix well. This is *DF 10^{-6}*.

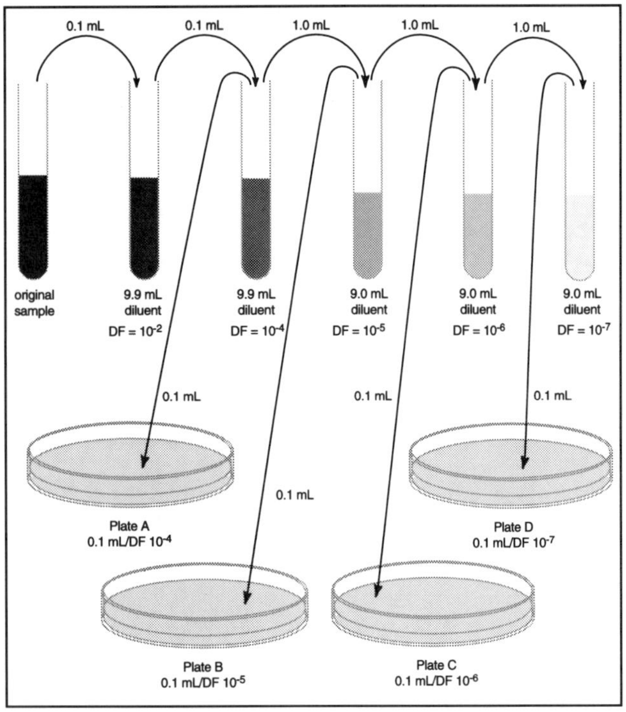

FIGURE 6-1a.

Serial Dilution *This is an illustration of the dilution scheme outlined in the test protocol. Use it as a guide while following the outlined procedure. As you perform the series of dilutions you will be assigning "dilution factors" to the tubes and their contents. These dilution factors do not indicate the concentration of cells in any of the tubes but how much the original sample has been diluted thus far. The number of colonies formed on the plates will provide the needed information to calculate the original cell concentration. Use the formulas provided in the Test Protocol section to do your calculations. For an explanation of dilution factors refer to the Precautions section.*

6. Aseptically transfer 1.0 mL DF 10^{-6} to a 9.0 mL dilution tube. Mix well. This is *DF 10^{-7}*.
7. Aseptically transfer 0.1 mL DF 10^{-4} to plate A. Using the spread plate technique disperse the diluent evenly over the entire surface of the agar. Repeat with the second plate A and label both plates "0.1 mL/DF 10^{-4}" (the volume transferred to the plate *and* the dilution factor of the tube).
8. Following the same procedure, transfer 0.1 mL volumes from DF 10^{-5}, DF 10^{-6} and DF 10^{-7} to plates B, C and D respectively. Label the plates accordingly.
9. Invert the plates and incubate at 35°C for 24 to 48 hours.
10. After incubation set the countable plates aside (plates with 30 to 300 colonies) and properly dispose of the uncountable plates. Only one pair of plates should be countable.
11. Count the colonies on both plates. Stab each colony with a toothpick as you record it with a hand tally counter. Calculate the average of both plates. (See Figure 6-1b.)
12. Using the formula below, determine the cell density of the original sample.

$$\text{Original cell density} = \left(\frac{\text{\# colonies}}{\text{mL plated}}\right)\left(\frac{1}{\text{dilution factor}}\right)$$

Original cell density =

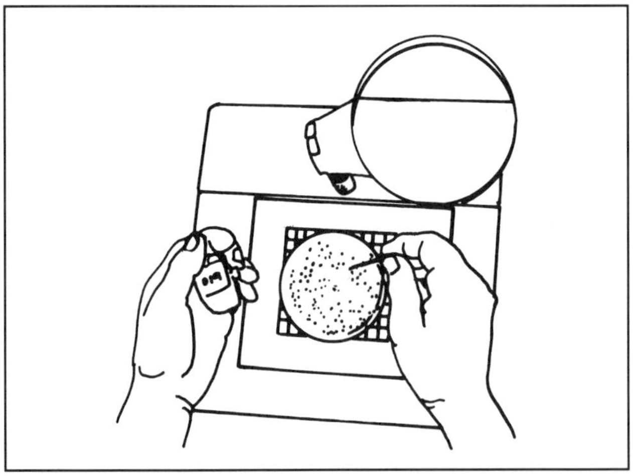

FIGURE 6-1b.

Counting Bacterial Colonies *Place the open plate on the colony counter, turn on the light and adjust the magnifying glass until all the colonies are visible. Using the grid in the background as a guide, count the colonies one section at a time. Mark each colony with a toothpick (to avoid counting it more than once) as you record with a hand tally counter.*

Precaution

⚠ Dilution factors in serial dilutions are calculated using the following formula: $DF_2 = (V_1)(DF_1)/V_2$ where DF_2 is the new dilution factor being calculated, DF_1 is the dilution factor of the sample being diluted (undiluted samples have a dilution factor of 1.0), V_1 is the volume of sample to be diluted and V_2 is the combined volume of sample and diluent. For example, to calculate the new dilution factor when adding 0.1 mL of DF 10^{-2} to 9.9 mL of diluent, use the formula thus:

$$DF_2 = \frac{(V_1)(DF_1)}{V_2}$$

$$DF_2 = \frac{(0.1\text{ mL})(10^{-2})}{10.0\text{ mL}} = 10^{-4}$$

For further explanation of dilution factors and serial dilutions, see Figure 6-1a on the previous page and Figure 6-1 on page 82 of the *Photographic Atlas*.

References

Collins, C. H., Patricia M. Lyne, J. M. Grange. 1995. Page 149 in *Collins and Lyne's Microbiological Methods, 7th Ed.* Butterworth-Heinemann, UK.

DIFCO Laboratories. 1984. Page 619 in *DIFCO Manual, 10th Ed.* DIFCO Laboratories, Detroit, MI.

Koch, Arthur L. 1994. Page 254 in *Methods for General and Molecular Bacteriology*, edited by Philipp Gerhardt, R. G. E. Murray, Willis A. Wood and Noel R. Krieg, American Society for Microbiology, Washington, DC.

Postgate, J. R. 1969. Page 611 in *Methods in Microbiology*, Vol. 1., edited by J. R. Norris and D. W. Ribbons, Academic Press, Inc., New York.

Direct Count

Photographic Atlas Reference

Page 84

Recipes

Staining/Diluting agents

Agent A

100% saturated crystal violet-ethanol solution	40.0 mL
NaCl	0.9 g
Distilled or deionized water	<60.0 mL*

Agent B

Ethanol	40.0 mL
NaCl	0.9 g
Distilled or deionized water	<60.0 mL*

*Add water to bring the total solution volume up to 100.0 mL.

Test Protocol

1. Transfer 0.1 mL from the original 24 hour culture tube to a non-sterile test tube.
2. Add 0.4 mL Agent A and 0.5 mL Agent B and mix well. (This dilution may not be suitable in all situations. Adjust the proportions of the broth culture and agents A and B if necessary to obtain a countable dilution. There should be at least two cells per small square to be a countable dilution.)
3. Place a drop of this mixture in the Petroff-Hausser counting chamber and cover with a coverslip.
4. Observe in the microscope and count the number of cells above each of five small squares.
5. Take the average and compute the original cell density using the formula below. Record your results in the space provided. (For further explanation of direct count calculations, refer to Precautions below and to page 84 of the *Photographic Atlas*.)

$$\text{Original cell density} = \left(\frac{\text{average \# cells / square}}{\text{volume above each square}}\right)\left(\frac{1}{\text{dilution factor}}\right)$$

$$\text{Original cell density} = \left(\frac{\text{average \# cells / square}}{5.0 \times 10^{-8}\ \frac{\text{ml}}{\text{square}}}\right)\left(\frac{1}{\text{dilution factor}}\right)$$

Original cell density =

Materials

Petroff-Hausser counting chamber with coverslip (See Figure 6-2)
Test tubes
1 mL pipettes with pipettor
Pasteur pipettes with bulbs
Hand counter
Staining agents A and B
24 hour broth culture of *Escherichia coli*

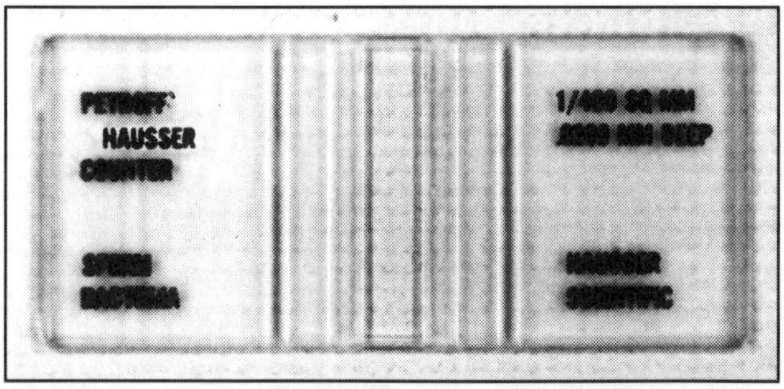

FIGURE 6-2.

Petroff-Hausser Counting Chamber *The Petroff-Hausser counting chamber is a device used for the direct counting of bacterial cells. To examine a bacterial broth or suspension, place a drop of the sample in the chamber, cover it with a coverslip and place it on the microscope stage. Locate the grid in the center using the low power objective lens. Do not increase the magnification until you have found the grid on low power. Increase the magnification, focusing one objective at a time, until you have the cells and the grid in focus with the oil immersion lens. Follow the instructions in the test protocol.*

Precaution

⚠ When calculating the original cell density, do not forget to include the dilution factor introduced with the addition of agents A and B. Where a *single* dilution is involved, dilution factor is calculated using this formula: DF = V_1/V_2 where V_1 is the volume of sample added and V_2 is the combined volume of sample and diluent. (For a discussion on calculating dilution factors in *serial* dilutions, refer to Viable Count on page 67)

The dilution factor for the dilution in #2 above is calculated thus:

$$DF = \frac{V_1}{V_2}$$

$$DF = \frac{0.1 \text{ mL}}{1.0 \text{ mL}} = 10^{-1}$$

References

Koch, Arthur L. 1994. Page 251 in *Methods for General and Molecular Bacteriology*, edited by Philipp Gerhardt, R. G. E. Murray, Willis A. Wood and Noel R. Krieg, American Society for Microbiology, Washington, DC.

Postgate, J. R. 1969. Page 611 in *Methods in Microbiology*, vol. 1., edited by J. R. Norris and D. W. Ribbons, Academic Press, Inc., New York.

Plaque Assay for Determination of Phage Titre

Photographic Atlas Reference

Page 86

Recipes

Nutrient Agar

Beef extract	3.0 g
Peptone	5.0 g
Agar	15.0 g
Distilled or deionized water	1.0 L
final pH = 6.8 ± 0.2 at 25°C	

Soft Agar

Beef extract	3.0 g
Peptone	5.0 g
Sodium chloride	5.0 g
Tryptone	2.5 g
Yeast extract	2.5 g
Agar	7.0 g
Distilled or deionized water	1.0 L

Dilution Tubes (0.9% NaCl)

NaCl	0.9 g
Distilled or deionized water	<100.0 mL*

*total solution volume is 100.0 mL

Medium Preparation

Nutrient Agar

1. Suspend, mix and boil the ingredients in one liter of distilled or deionized water until completely dissolved.
2. Cover loosely and sterilize in the autoclave at 15 lbs. pressure (121°C) for 15 minutes.
3. Remove from the autoclave and cool slightly.
4. Aseptically pour into sterile Petri dishes (15 mL/plate) and allow to cool to room temperature.

Soft Agar

1. Suspend, mix and boil the ingredients in one liter of distilled or deionized water until completely dissolved.
2. Transfer 2.5 mL portions to test tubes and cap loosely.
3. Sterilize in the autoclave at 15 lbs. pressure (121°C) for 15 minutes.
4. Remove from the autoclave and place in a hot water bath set at 45°C. Allow 30 minutes for the agar temperature to equilibrate.

Dilution Tubes

1. Dissolve the NaCl in approximately 90 mL of water.
2. Add water to bring the total volume up to 100.0 mL.
3. Cover loosely and sterilize in the autoclave at 121°C for 15 minutes.
4. Allow to cool to room temperature.
5. Aseptically dispense into sterile test tubes in the amounts of 9.0 mL and 9.9 mL.

Materials

1 mL sterile pipettes
Six 9.0 mL and one 9.9 mL dilution tubes
Six nutrient agar plates
Six soft agar tubes
Hot water bath set at 45°C
T4 coliphage (Carolina Biological #K3-12-4330)
24 hour broth culture of *Escherichia coli B* (T-series phage host — Carolina Biological #K3-12-4300))

Test Protocol

This procedure presumes the range of the original phage titer to be between 1.0×10^6 and 1.0×10^{11} PFU (Plaque Forming Units) /mL. If the sample you receive is not in this range, you will need to adjust your dilution scheme up or down, accordingly, to ensure obtaining countable plates and to avoid wasting media on unnecessary dilutions. (For an explanation of dilution factors and serial dilutions see Precautions below and Figure 6-1a in Viable Count.)

1. Aseptically transfer 0.1 mL of the original sample to a 9.9 mL dilution tube. Mix well. This is *DF* (dilution factor) 10^{-2}.
2. Aseptically transfer 1.0 mL DF 10^{-2} to a 9.0 mL dilution tube. Mix well. This is *DF* 10^{-3}.
3. Aseptically transfer 1.0 mL DF 10^{-3} to a 9.0 mL dilution tube. Mix well. This is *DF* 10^{-4}.
4. Continue in this manner (adding 1.0 mL to 9.0 mL) until you have completed DF 10^{-8}.

5. Remove one soft agar tube from the hot water bath and add 0.1 mL DF 10^{-3} and 0.3 mL of the *E. coli* broth culture. Mix well and immediately pour onto a nutrient agar plate. Gently tilt the plate back and forth until the soft agar mixture is spread evenly across the solid medium. Label the plate 0.1 mL/DF 10^{-3} (the volume of phage transferred to the plate *and* the dilution factor of the diluent tube).
6. Repeat this procedure with dilutions 10^{-4}, 10^{-5}, 10^{-6}, 10^{-7}, and 10^{-8}. Label the plates accordingly.
7. Allow the agar to solidify completely.
8. Invert the plates and incubate aerobically at 35°C for 24 to 48 hours.
9. After incubation, count the plaques (Figure 6-3).
10. Determine the original phage titer using the formula below. Record your results in the space provided.

$$\text{Original phage density} = \left(\frac{\text{\# plaques counted}}{\text{mL virus plated}}\right)\left(\frac{1}{\text{dilution factor}}\right)$$

Original phage density =

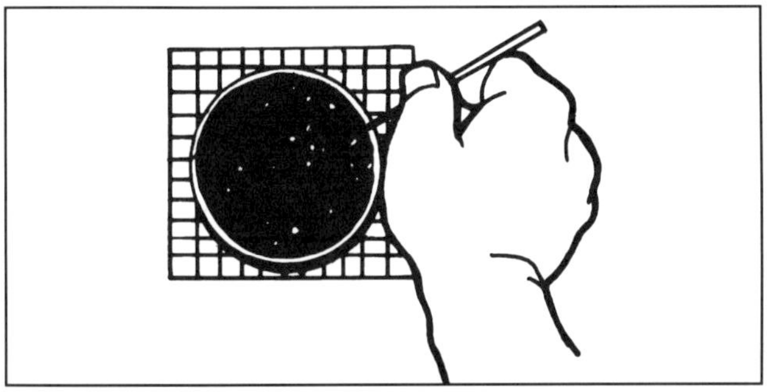

FIGURE 6-3.

Counting Plaques *Place the inverted plate on the colony counter, turn on the light and adjust the magnifying glass until all the plaques are visible. Using the grid in the background as a guide, count the plaques one section at a time. Mark each plaque with a felt tip marker (to avoid counting it more than once) as you record it with a hand tally counter.*

Precautions

⚠ The soft agar tubes will solidify rather quickly upon removal from the hot water bath. Be prepared to use them immediately.

⚠ Prewarming the nutrient agar plates in the incubator will slow down the solidifying of the soft agar and allow more time for it to spread evenly over the solid medium.

⚠ Dilution factors in serial dilutions are calculated using the following formula: $DF_2 = (V_1)(DF_1)/V_2$ where DF_2 is the new dilution factor being calculated, DF_1 is the dilution factor of the sample being diluted (undiluted samples have a dilution factor of 1.0), V_1 is the volume of sample to be diluted and V_2 is the combined volume of sample and diluent. For example, to calculate the new dilution factor when adding 0.1 mL of DF 10^{-2} to 9.9 mL of diluent, use the formula thus:

$$DF_2 = \frac{(V_1)\,(DF_1)}{V_2}$$

$$DF_2 = \frac{(0.1\text{ mL})\,(10^{-2})}{10.0\text{ mL}} = 10^{-4}$$

References

Collins, C. H., Patricia M. Lyne, J. M. Grange. 1995. Page 149 in *Collins and Lyne's Microbiological Methods, 7th Ed.* Butterworth-Heinemann, UK.

DIFCO Laboratories. 1984. Page 619 in *DIFCO Manual, 10th Ed.* DIFCO Laboratories, Detroit, MI.

Province, David L. and Roy Curtiss III. 1994. Page 328 in *Methods for General and Molecular Bacteriology*, edited by Philipp Gerhardt, R. G. E. Murray, Willis A. Wood and Noel R. Krieg, American Society for Microbiology, Washington, DC.

SECTION SEVEN

Medical, Food and Environmental Microbiology

Ames Test

Photographic Atlas Reference

Page 89

Recipes

Nutrient Broth + 0.5% NaCl

Beef extract	0.6 g
Peptone	1.0 g
NaCl	1.0 g
Distilled or deionized water	200.0 mL

Complete Medium (CM)

Beef extract	3.0 g
Peptone	5.0 g
Sodium chloride	5.0 g
Agar	20.0 g
Distilled or deionized water	1.0 L

Minimal Medium (MM)

Dextrose (glucose)	20.0 g
50x Vogel-Bonner salts	20.0 mL
Histidine	0.00016 g
Biotin	0.00025 g
Agar	20.0 g
Distilled or deionized water	1.0 L

50x Vogel-Bonner Salts

Magnesium sulfate	10.0 g
Citric acid	100.0 g
Dipotassium phosphate	500.0 g
Monosodium ammonium phosphate	175.0 g
Distilled or deionized water	<1.0 liter*

*total solution volume is 1.0 liter (See Precautions below)

Medium Preparation

Nutrient Broth + 0.5% NaCl

1. Suspend, mix and warm the ingredients in 200.0 mL of distilled or deionized water until dissolved completely.
2. Dispense 10.0 mL volumes into 2 test tubes and 90.0 mL volumes into two dilution bottles.
3. Sterilize in the autoclave at 121°C for 15 minutes.
4. Remove from the autoclave and allow to cool.

Complete Medium

1. Suspend, mix and boil the ingredients in one liter of distilled or deionized water until completely dissolved.

2. Cover loosely and sterilize in the autoclave at 15 lbs. pressure (121°C) for 15 minutes.
3. Remove from the autoclave and cool slightly.
4. Aseptically pour into sterile petri dishes (15 mL/plate) and allow to cool to room temperature.

Minimal Medium

1. Add 1.6 mg histidine to 10.0 mL distilled or deionized water and filter sterilize.
2. Add 2.5 mg biotin to 10.0 mL distilled or deionized water and filter sterilize.
3. Prepare the 50x Vogel-Bonner salts solution by adding the ingredients to *just enough* water to dissolve them while heating and stirring. After the ingredients are dissolved add enough water to bring the total volume up to exactly one liter.
4. Suspend, mix and boil the agar in 500.0 mL of distilled or deionized water until completely dissolved.
5. Suspend and mix the dextrose in 500.0 mL of distilled or deionized water until completely dissolved.
6. Cover the agar and dextrose containers loosely and sterilize in the autoclave at 121°C for 15 minutes.
7. Remove from the autoclave and allow to cool to 80°C.
8. Aseptically add 1.0 mL histidine solution, 1.0 mL biotin solution, and 20 mL 50x Vogel-Bonner salts to the glucose solution and mix well.
9. Add the glucose solution to the agar solution, mix well and aseptically pour into sterile Petri dishes (15 mL/plate).

Materials

- Four MM plates
- Four CM plates
- Two tubes containing 10.0 mL nutrient broth + 0.5% NaCl
- Two bottles containing 90.0 mL nutrient broth + 0.5% NaCl
- Centrifuge
- Sterile centrifuge tubes
- Small beaker containing alcohol and forceps
- Bottle of 1x Vogel-Bonner salts (To make 1x Vogel-Bonner salts, add 1.0 mL 50x Vogel-Bonner solution to 49.0 mL water.)
- Sterile filter discs made with a paper punch
- Two sterile petri dishes (for soaking filter paper discs)
- Sterile 10 mL pipettes
- Container for disposal of supernatant (to be autoclaved)
- DMSO
- Test substance (any substance which has possible mutagenic properties and does not contain histidine or protein)
- Broth culture of *Salmonella typhimurium TA 1535* *
- Broth culture of *Salmonella typhimurium TA 1538* *

*The cultures used for this exercise must be prepared as follows:

1. 24 hours before the test, inoculate the two 10.0 mL broth tubes with TA 1535 and TA 1538 and incubate at 35°C together with the two sterile 90.0 mL broths.
2. 5½ hours before the test, pour the TA 1538 culture into one of the sterile 90.0 mL broths and return it to the incubator until time for the test.
3. 4 hours before the test, pour the TA 1535 culture into the other 90.0 mL sterile broth and return it to the incubator until time for the test.

Test Protocol

1. Soak 4 filter paper discs in DMSO and 4 filter paper discs in the test substance.
2. Pipette 10.0 mL TA 1535 into a sterile centrifuge tube. Do the same with TA 1538, then label the tubes.
3. Centrifuge the tubes on high speed for 10 minutes.
4. Being careful not to disturb the cell pellet at the bottom, decant the supernatant from each centrifuge tube. (See Figure 7-1a.)
5. Resuspend the cell pellets by adding 1.0 mL sterile 1x Vogel-Bonner salts to each tube and mixing well.

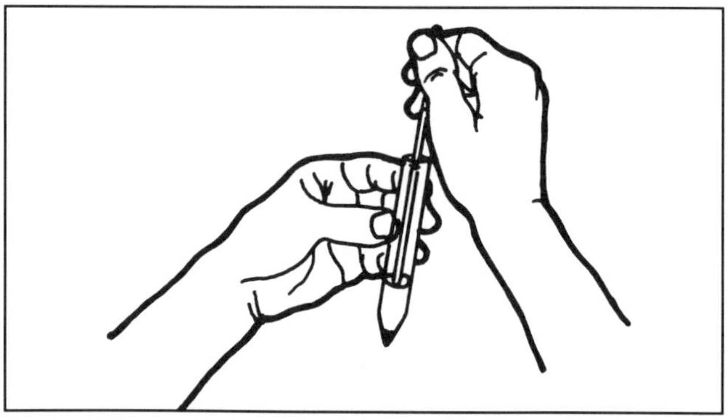

FIGURE 7-1A.

Decanting *This variation of decanting uses a Pasteur pipette and bulb to draw the supernatant away from the pellet. Discharge the air from the bulb before placing it in the tube, then slowly lower the pipette into the fluid and draw off the supernatant. Repeat this procedure until all of the supernatant has been removed. A little bit of the pellet may also be removed, but this can be minimized by working slowly to avoid agitation of the fluid and cells. Dispose of the supernatant in a container to be autoclaved.*

6. Using spread plate technique, spread 0.1 mL of the resuspended TA 1535 on each of two MM plates and two CM plates. Do the same with TA 1538.
7. Sterilize the forceps by passing them through the Bunsen burner flame and allowing the alcohol to burn off.
8. Using the sterile forceps, place the discs in the centers of the plates as follows:

	CM with TA 1535	CM with TA 1538	MM with TA 1535	MM with TA 1538
DMSO	●	●	●	●
Test Substance	○	○	○	○

9. Incubate aerobically at 35°C for 48 hours.
10. Measure and compare the zones of inhibition on the CM plates (Figure 7-1b). Count the colonies on the MM plates and compare (Figure 7-1c) .
11. Record your results in the table below.

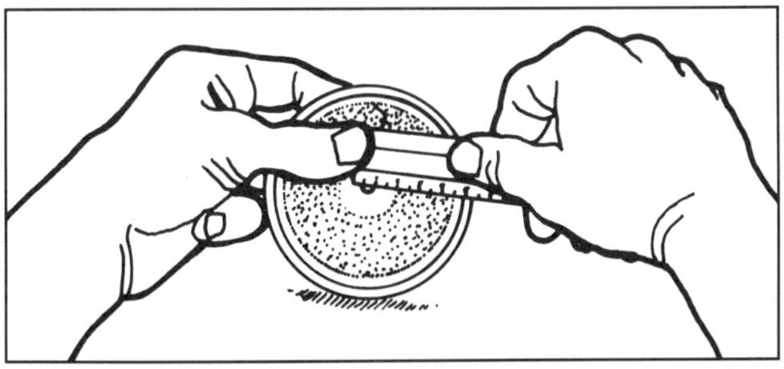

FIGURE 7-1b.

Measuring the Zone of Inhibition *Using a metric ruler, measure the shortest distance from the edge of the paper disc to the perimeter of the clearing. Do this with all the CM plates and record your measurement in mm.*

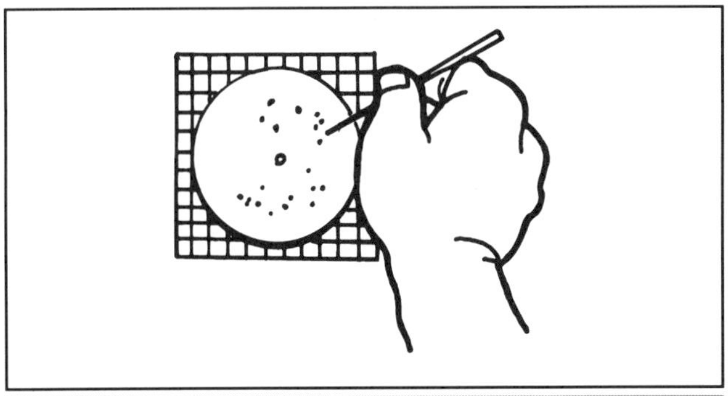

FIGURE 7-1c.

Counting The Mutant Colonies *Place the open plate on the colony counter, turn on the light and adjust the magnifying glass until all the colonies are visible. Using the grid in the background as a guide, count the colonies one section at a time. Mark each colony with a toothpick (to avoid counting it more than once) as you keep track of the number using a hand tally counter. Do this with all of the MM plates and record your results.*

Precautions

⚠ Be sure to load the centrifuge evenly and watch it at all times when it is running.

⚠ While preparing the 50x Vogel-Bonner salts be careful not to add too much water initially or the addition of the dry ingredients will increase the volume to greater than one liter.

References

Eisenstadt, Bruce C. Carlton, and Barbara J. Brown. 1994. Page 311 in *Methods for General and Molecular Bacteriology*, edited by Philipp Gerhardt, R. G. E. Murray, Willis A. Wood and Noel R. Krieg, American Society for Microbiology, Washington, DC.

Maron, D. M. and B. N. Ames. 1983. *Mutation Research*, 113:173-215.

	ZONE DIAMETER		COLONIES COUNTED	
	CM with TA 1535	CM with TA 1538	MM with TA 1535	MM with TA 1538
DMSO				
Test Substance				

Antibiotic Sensitivity — Kirby-Bauer Test

Photographic Atlas Reference

Page 91

Recipe

Mueller-Hinton II Agar

Beef extract	2.0 g
Acid hydrolysate of casein	17.5 g
Starch	1.5 g
Agar	17.0 g
Distilled or deionized water	1.0 L

final pH = 7.3 ± 0.1 at 25°C

Medium Preparation

1. Suspend the ingredients in one liter of distilled or deionized water, mix well and boil to dissolve completely.
2. Cover loosely and sterilize in the autoclave at 121°C (15 lbs.) for 15 minutes.
3. Remove from the autoclave, allow to cool slightly.
4. Aseptically pour into sterile Petri dishes to a depth of 4 mm.
5. Allow to cool to room temperature.

Materials

Four Mueller-Hinton agar plates
Commercially prepared discs impregnated with the antibiotics: streptomycin, tetracycline, penicillin, chloramphenicol, sulfisoxazole and trimethoprim.
Antibiotic disc dispenser (See the dispenser in Figure 7-2a. If you don't have one, you can use the forceps and alcohol listed below.)
Sterile cotton swabs
Metric ruler
Zone diameter interpretive table published by the National Committee for Clinical Laboratory Standards (NCCLS). Refer to page 92 of the *Photographic Atlas* for more information.
Small beaker with alcohol and forceps
24 hour pure broth cultures of:
Escherichia coli
Staphylococcus aureus

Test Protocol

1. Dip a sterile swab into a broth culture and wipe off the excess on the inside of the tube.
2. Inoculate a Mueller-Hinton plate by smearing the entire surface of the agar with the swab. Inoculate two plates with each culture being tested. Be sure to use a different swab for each inoculation.
3. Label the plates with the organisms' names, your name and the date.
4. Apply the streptomycin, tetracycline, penicillin and chloramphenicol discs to the agar surface of one plate for each organism. You can apply the discs either singly using sterile forceps, or with a dispenser (Figure 7-2a). The forceps can be sterilized by passing them through the Bunsen burner flame and allowing the alcohol to burn off. Be sure to space the discs sufficiently (4 to 5 cm) to prevent overlapping zones of inhibition.
5. Press each disc gently with sterile forceps so that it makes good contact with the agar surface.
6. Using sterile forceps (as in #4 above) apply the sulfisoxazole and trimethoprim discs to the agar

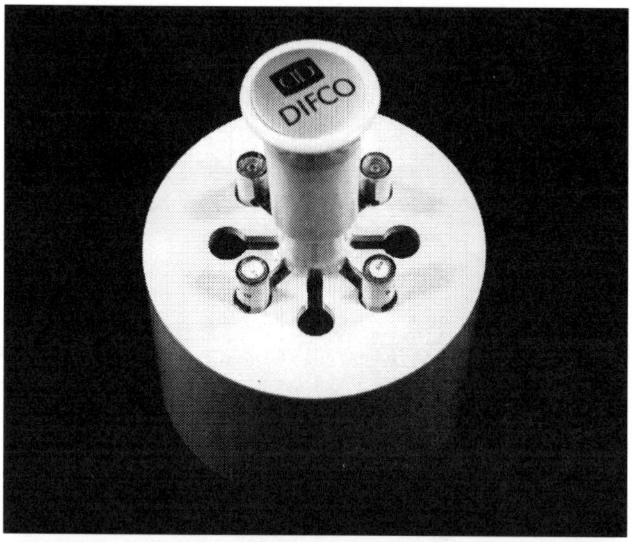

FIGURE 7-2a.

Dispensing the Antibiotic Discs *Center the dispenser over the open plate and press firmly* only once *to deposit the discs. Gently press each disc with sterile forceps to make sure it is making good contact with the agar surface. Cover the plate, invert it and place it in the incubator.*

surface of the two remaining plates in the following manner:

a. On the *E. coli* plate place the discs exactly 25 mm apart.

b. On the *S. aureus* plate place the discs exactly 29 mm apart.

7. Invert the plates and incubate them aerobically at 35°C for 18 hours.
8. Remove the plates from the incubator and measure the zones of inhibition. (See Figure 7-2b.)
9. Record your results in the table below.

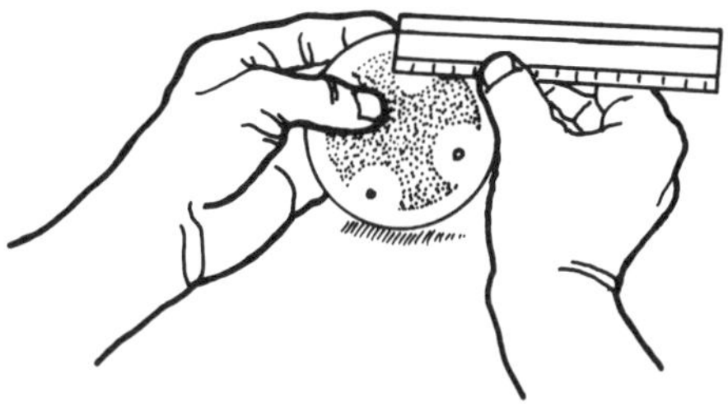

FIGURE 7-2b.

Measuring the Antimicrobial Susceptibility Zones
Place the inverted plate on the colony counter and, using a metric ruler, measure the entire diameter of each clearing. To get the most accurate reading, place the 10 mm mark on left edge of the zone, measure the diameter and subtract 10 mm from the measurement. Record your results and dispose of the plates properly.

Precaution

⚠ Take care to place the trimethoprim and sulfisoxazole discs exactly the recommended distances apart.

References

Collins, C. H., Patricia M. Lyne, J. M. Grange. 1995. Page 128 in *Collins and Lyne's Microbiological Methods, 7th Ed.* Butterworth-Heinemann, UK.

Jorgensen, James H. et. al. 1994. *Performance Standards for Antimicrobial Susceptibility Testing; Fifth Informational Supplement, Vol. 14 no. 16*, NCCLS, Villanova, PA.

Power, David A. and Peggy J. McCuen. 1988. Page 204 in *Manual of BBL® Products and Laboratory Procedures, 6th Ed.* Becton Dickinson Microbiology Systems, Cockeysville, MD.

Woods, Gail L. and John A. Washington. 1995. Page 1337 in *Manual of Cinical. Microbiology, 6th Ed.*, edited by Patrick R. Murray, Ellen Jo Baron, Michael A. Pfaller, Fred C. Tenover and Robert H. Yolken, ASM Press, Washington, DC.

ORGANISM	ZONE DIAMETER			
	Streptomycin	Tetracycline	Penicillin	Chloramphenicol
Escherichia coli				
Staphylococcus aureus				

Membrane Filter Technique

Photographic Atlas Reference

Page 93

Recipe

Levine Eosin Methylene Blue (EMB) Agar

Peptone	10.0 g
Lactose	10.0 g
Dipotassium phosphate	2.0 g
Agar	15.0 g
Eosin Y	0.4 g
Methylene blue	0.065 g
Distilled or deionized water	1.0 L

final pH = 7.1 ± 0.2 at 25°C

Medium Preparation

1. Suspend the ingredients in one liter of distilled or deionized water, mix well and boil to dissolve completely.
2. Cover loosely and sterilize in the autoclave at 121°C (15 lbs.) for 15 minutes.
3. Remove from the autoclave and allow to cool slightly.
4. Aseptically pour into sterile Petri dishes (15 mL/plate).
5. Allow to cool to room temperature.

Test Protocol

1. Sterilize the forceps by placing them in the Bunsen burner flame long enough to ignite the alcohol. Once the forceps are sterile, use them to place the filter (grid facing up) between the two halves of the filter housing. Clamp all parts together.
2. Insert the filter housing into the suction flask as shown in Figure 7-3a.
3. Pour the water sample into the filter housing funnel and begin the suction.
4. When the water has passed through the membrane filter, stop the suction.
5. Sterilize the forceps again by flaming and carefully remove the filter.
6. Place the filter on the EMB agar being careful not to fold it or create air pockets under it (see Figure 7-3b).
7. Invert the plate and incubate it aerobically at 35°C for 24 hours.
8. Remove the plate and count the dark purple colonies and the colonies which produce a green metallic sheen. If there are none, incubate the plate another 24 hours before scoring it as negative.
9. Use the formula on page 93 of the *Photographic Atlas* to determine the water sample's potability.

Precaution

⚠ Place the vacuum flasks in test tube baskets or otherwise secure them to the table to prevent the tubing from tipping them over.

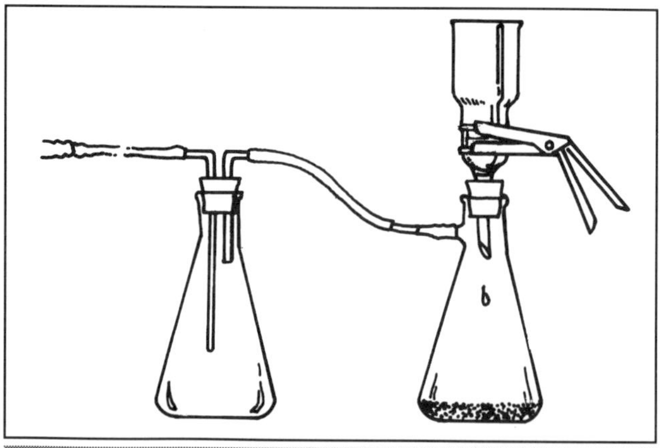

FIGURE 7-3a.

Membrane Filter Apparatus *Assemble the membrane filter apparatus as shown in this photograph. It is important to use two suction flasks to avoid getting water into the vacuum source. Secure the flasks on the table as the tubing will make them top heavy.*

Materials

- One EMB plate
- One sterile membrane filter (pore size 0.45 µm)
- Sterile membrane filter suction apparatus (see Figure 7-3a.)
- 100.0 mL water sample taken from any source where potability may be questionable
- Small beaker containing alcohol and forceps
- Vacuum source (This can be either a vacuum pump or an aspirator connected to a faucet. Make sure that any faucet used is equipped with a properly installed anti-siphon device.)

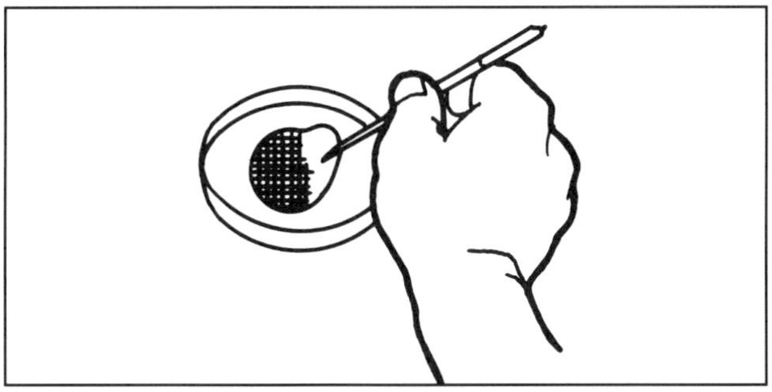

FIGURE 7-3b.

Placing the Filter On the Agar Plate *Using sterile forceps carefully place the filter onto the agar surface with the grid facing up. Try not to allow any air pockets under the filter since contact with the agar surface is essential for bacterial growth. Allow a few minutes for the filter to adhere to the agar before inverting the plate.*

References

Chan, E. C. S.,Pelczar, Jr., Krieg Noel R. 1986. Page 291 in *Laboratory Exercises In Microbiology*. McGraw-Hill Book Company.

Collins, C. H., Patricia M. Lyne, J. M. Grange. 1995. Page 270 in *Collins and Lyne's Microbiological Methods, 7th Ed.* Butterworth-Heinemann, UK.

DIFCO Laboratories. 1984. Page 515 in *DIFCO Manual, 10th Ed.* DIFCO Laboratories, Detroit, MI.

Mulvany, J. G. 1969. Page 205 in *Methods in Microbiology, Vol. 1,* edited by J. R. Norris and D. W. Ribbons, Academic Press Inc., New York.

Power, David A. and Peggy J. McCuen. 1988. Page 153 in *Manual of BBL® Products and Laboratory Procedures, 6th Ed.* Becton Dickinson Microbiology Systems, Cockeysville, MD.

Methylene Blue Reductase Test

Photographic Atlas Reference

Page 95

Recipe

Methylene Blue Solution

Methylene blue dye	8.8 mg
Distilled or deionized water	200.0 mL

Test Protocol

1. Inoculate tube A with *Escherichia coli.* (The addition of *E. coli* to the milk simulates milk of poor quality. However, any milk of questionable quality can be tested with this procedure. To do this, skip step #1 and begin at step #2. Use as many samples as you like, but be sure to label them clearly.)
2. Aseptically add 1.0 mL methylene blue solution to test tube A and to test tube B. Cap the tubes tightly and invert several times to mix thoroughly.
3. Place tubes A and B in the hot water bath and note the time.
4. Place the control tube in the refrigerator.
5. After 5 minutes of incubation remove the tubes, invert them *once* to mix again, then return them to the water bath. Check the time on the clock and record it in the table below under "STARTING TIME".
6. Using the control tube for color comparison, check tubes A and B at 30 minute intervals and record the time when each becomes white. Poor quality milk takes less than 2 hours; good quality milk takes longer than 6 hours.
7. Using the table below, calculate the time it takes for each milk sample to become white.

Materials

Three sterile test tubes each containing 10.0 mL of fresh milk, labeled "A", "B" and "Control"
Sterile 1.0 mL pipettes
Hot water bath set at 35°C
Methylene Blue solution
24 hour pure culture on solid medium:
Escherichia coli
Clock or wristwatch

Precaution

⚠ Be sure to record the *clock* times in the table under "STARTING TIME" and "ENDING TIME". The "ELAPSED TIME" is the difference between the two clock times and is the actual number of hours or minutes taken to complete the reaction.

References

Bailey, R. W., and E. G. Scott. 1966. Page 114 and 306 in *Diagnostic Microbiology*, 2nd Ed., C. V. Mosby Company, St. Louis, MO.

Benathen, Isaiah. 1993. Page 132 in *Microbiology With Health Care Applications*, Star Publishing Company, Belmont, CA.

Power, David A. and Peggy J. McCuen. 1988. Page 62 in *Manual of BBL® Products and Laboratory Procedures*, 6th Ed. Becton Dickinson Microbiology Systems, Cockeysville, Md.

Richardson (ed.). 1985. *Standard Methods for the Examination of Dairy Products*, 15th Ed. American Public Health Association, Washington DC.

TUBE	STARTING TIME T_s	ENDING TIME T_e	ELAPSED TIME ($T_e - T_s$)
A			
B			

Snyder Test

Photographic Atlas Reference

Page 96

Recipe

Snyder Agar

Pancreatic digest of casein	13.5 g
Yeast extract	6.5 g
Dextrose	20.0 g
Sodium chloride	5.0 g
Agar	16.0 g
Bromcresol green	0.02 g
Distilled or deionized water	1.0 L

final pH = 4.8 ± 0.2 at 25°C

Medium Preparation

1. Suspend the ingredients in one liter distilled or deionized water, mix well and boil to dissolve completely.
2. Transfer 7.0 mL portions to test tubes and cap loosely.
3. Sterilize in the autoclave at 118–121°C for 15 minutes.
4. Remove from the autoclave and place in a hot water bath set at 45–50°C. Allow at least 30 minutes for the agar temperature to equilibrate before beginning the exercise.

Materials

Hot water bath set at 45–50°C
Small sterile beakers
Sterile 1 mL pipettes with bulbs
Two Snyder agar tubes

Test Protocol

1. Collect a small sample of saliva (about 0.5 mL) in the sterile beaker.
2. Aseptically add 0.1 mL of the sample to a Snyder agar tube and roll the tube between your hands until the saliva is uniformly distributed throughout the agar.
3. Allow the agar to cool to room temperature; do not slant.
4. Incubate with an uninoculated control at 35°C for up to 72 hours.
5. Check the tubes at 24 hour intervals for yellow color formation.
6. Record your results and determine your susceptibility to tooth decay using the information on page 96 of the *Photographic Atlas*.

References

DIFCO Laboratories. 1984. Page 619 in *DIFCO Manual, 10th Ed.* DIFCO Laboratories, Detroit, MI.

Power, David A. and Peggy J. McCuen. 1988. Page 247 in *Manual of BBL® Products and Laboratory Procedures, 6th Ed.* Becton Dickinson Microbiology Systems, Cockeysville, MD.

	24 HRS.	48 HRS.	72 HRS.
COLOR			

SECTION EIGHT

Host Defenses, Immunology and Serology

Differential Blood Cell Count

Photographic Atlas Reference

Pages 97

Test Protocol

1. Obtain a blood smear slide and locate a field where the cells are spaced far enough apart to allow easy counting. (The cells should be fairly dense on the slide, but not overlapping.)
2. Using the oil immersion lens, scan the slide using the pattern shown in Figure 8-1.
3. Make a tally mark in the appropriate box of the table for the first 100 leukocytes you see.
4. Calculate percentages and compare your results with the accepted normal values.
5. Repeat with a pathological blood smear (if available).

Materials

Commercially prepared human blood smear slides (Wright's or Giemsa stain)

Optional: Commercially prepared abnormal human blood smear slides (*e.g.* infectious mononucleosis, eosinophilia, or neutrophilia)

Precautions

⚠ Remember that a microscope image is inverted. If you want the image to move left, you must move the slide to the right.

⚠ Be careful not to overlap fields when scanning the specimen. Choose a "landmark" blood cell at the right side of the field and move the slide horizontally until that cell disappears off the left side.

⚠ Avoid diagonal movement of the slide. As you scan, use the mechanical stage knobs separately to move the slide up and back or to the right in straight lines.

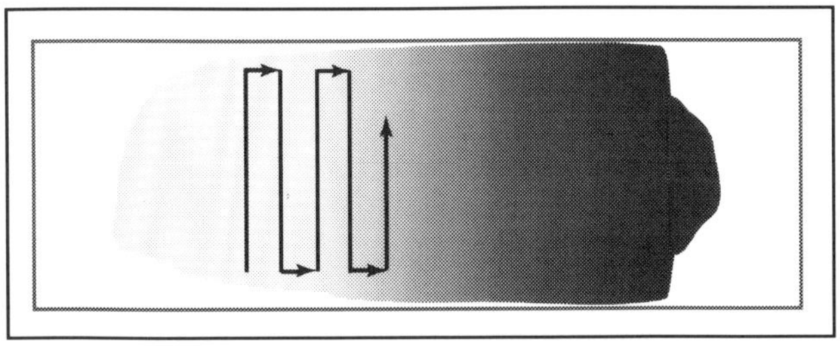

FIGURE 8-1.

Follow a Systematic Path *A systematic scanning path is used to avoid wandering around the slide and perhaps counting some cells more than once.*

NORMAL BLOOD

	MONOCYTES	LYMPHOCYTES	SEGMENTED NEUTROPHILS	BAND NEUTROPHILS	EOSINOPHILS	BASOPHILS
Number						
Percentage						
Expected Percentage	3 – 7%	25 – 33%	55 – 65% (All neutrophils)		1 – 3%	0.5 – 1%

ABNORMAL BLOOD

	MONOCYTES	LYMPHOCYTES	SEGMENTED NEUTROPHILS	BAND NEUTROPHILS	EOSINOPHILS	BASOPHILS
Number						
Percentage						
Expected Percentage	3 – 7%	25 – 33%	55 – 65% (All neutrophils)		1 – 3%	0.5 – 1%

References

Brown, Barbara A. 1993. *Hematology — Principles and Procedures, 6th Ed.* Lea and Febiger, Philadelphia, PA.

Eroschenko, Victor P. 1993. *di Fiore's Atlas of Histology with Functional Correlations, 7th Ed.* Lea and Febiger, Philadelphia, PA.

Junqueira, L. Carlos, Jose Carneiro and Robert O. Kelley. 1995. *Basic Histology, 8th Ed.* Appleton & Lange, Norwalk, CT.

Other Immune Cells and Organs

Photographic Atlas Reference

Page 99

Procedure

Examine the slides. On each, identify distinguishing characteristics and components relevant to the immune system.

1. Loose areolar tissue: mast cells, macrophages, collagen and elastic fibers
2. Liver: Kupffer cells, hepatic plates, sinusoids, central veins

Materials

Prepared slides of:
- loose areolar tissue
- liver (stained for Kupffer cells)
- lymph node
- thymus
- tonsil
- spleen
- lung
- ileum

3. Lymph node: lymph follicles (nodules) with germinal centers, lymphocytes, macrophages in sinuses (possibly), capsule
4. Thymus: lobules, lymphocytes, thymic corpuscles, capsule
5. Tonsil: crypts, lymph follicles (nodules), lymphocytes
6. Spleen: capsule, lymph nodules (white pulp) with central artery, blood sinusoids (red pulp)
7. Lung: lymph nodules lining respiratory tree, lymphocytes
8. Ileum: Peyer's patch with lymph nodules, lymphocytes

Precaution

⚠ Regions of lymphatic tissue are identifiable even at low power. Use higher magnification for examining detail.

References

Eroschenko, Victor P. 1993. *di Fiore's Atlas of Histology with Functional Correlations, 7th Ed.* Lea and Febiger, Philadelphia, PA.

Junqueira, L. Carlos, Jose Carneiro and Robert O. Kelley. 1995. *Basic Histology, 8th Ed.* Appleton & Lange, Norwalk, CT.

Precipitation Reactions — Precipitin Ring

Photographic Atlas Reference

Page 102

Materials

One clean Durham tube
Horse serum
Horse albumin antiserum
Pasteur pipettes

Test Protocol

1. Using a clean Pasteur pipette, carefully add horse albumin antiserum to the Durham tube. Fill from the bottom of the tube until it is about 1/3 full. (See Figure 8-2a.)
2 Using a different Pasteur pipette, add the horse serum in such a way that a sharp and distinct second layer is formed without any mixing of the two solutions. (See Figure 8-2b.)
3. Incubate at 35°C for one hour.
4. Observe the tube for the characteristic ring formed at the equivalence zone. If after one hour there is no ring, place the tube in the refrigerator for 12 to 24 hours and then recheck.

Precaution

⚠ It is critical that the serum be *very* carefully placed on top of the antiserum *without any mixing* of the two solutions. Success can usually be achieved by allowing the serum to slowly trickle down the inside of the glass.

Reference

Lam, Joseph S. and Lucy M. Mutharia. 1994. Page 120 in *Methods for General and Molecular Bacteriology*, edited by Philipp Gerhardt, R. G. E. Murray, Willis A. Wood and Noel R. Krieg, American Society for Microbiology, Washington, DC.

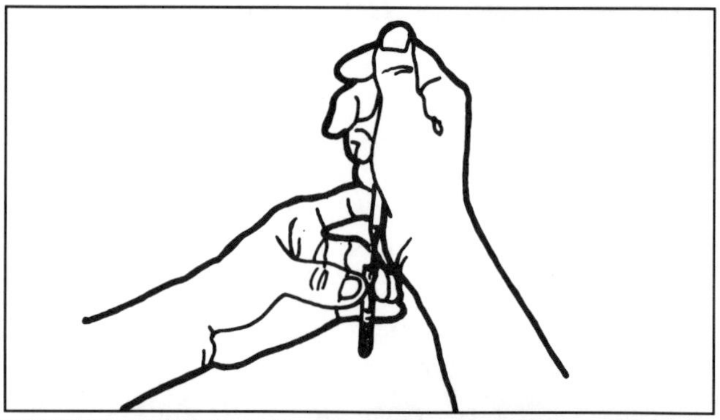

FIGURE 8-2a.

Placing the Antibody In the Tube *When adding the antibody to the tube, fill from the bottom to avoid creating air bubbles. Fill about 1/3 full.*

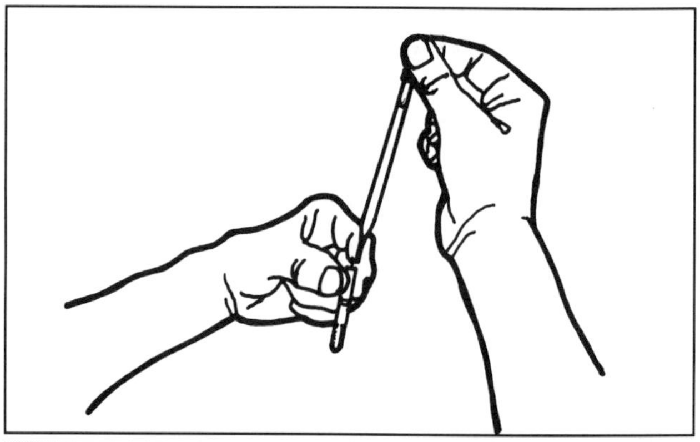

FIGURE 8-2b.

Layering the Antigen On Top of the Antibody *When adding the antigen layer to the tube containing antibody, it is essential that there be no mixing of the two solutions. Place the pipette containing the antigen into the tube about 1/2 cm from the antibody and let it slowly trickle down the inside of the glass. Allow to stand for one hour undisturbed.*

Precipitation Reactions — Double-Gel Immunodiffusion

Photographic Atlas Reference

Page 102

Recipe

Saline Agar

Sodium chloride	10.0 g
Agar	20.0 g
Distilled or deionized water	1.0 L

Medium Preparation

1. Suspend the ingredients in one liter distilled or deionized water, mix well and boil to dissolve completely.
2. Pour into Petri dishes to a depth of 3 mm. Do not replace the lids until the agar has solidified and cooled to room temperature.

Test Protocol

1. Center the saline agar plate over the template in Figure 8-3a.
2. Using the punch or glass dropper, cut the 7 wells in the agar as shown in Figure 8-3b. If the small agar discs don't come out with the dropper, use suction with the dropper and bulb to dislodge and remove them.
3. Number the peripheral wells #1 through #6 as shown in Figure 8-3a.
4. In a small test tube or mixing cup prepare a 1:1 mixture of anti-horse albumin and anti-bovine albumin.

Materials

- One saline agar plate
- 3 mm punch (a glass dropper with a 3 mm diameter tip will work)
- Template for cutting wells (See Figure 8-3a.)
- Anti-bovine albumin
- Anti-horse albumin
- Physiological saline
- 10% bovine serum (Prepared by adding 9 drops physiological saline to 1 drop 100% serum)
- 10% horse serum (Prepared by adding 9 drops physiological saline to 1 drop 100% serum)
- Disposable micropipettes

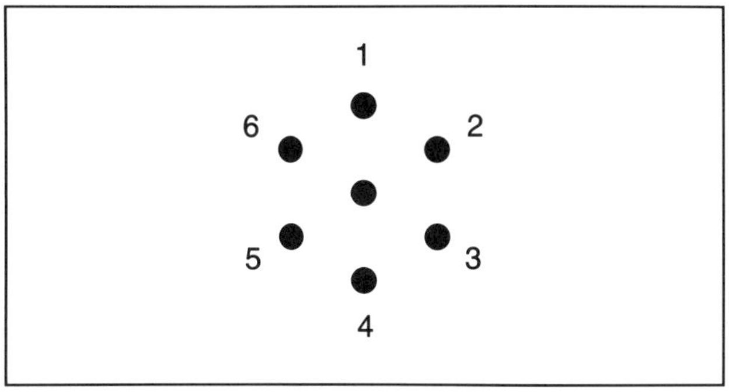

FIGURE 8-3a.

Double-Gel Immunodiffusion Well Template *Use this template as a guide when boring the wells in the saline agar plate.*

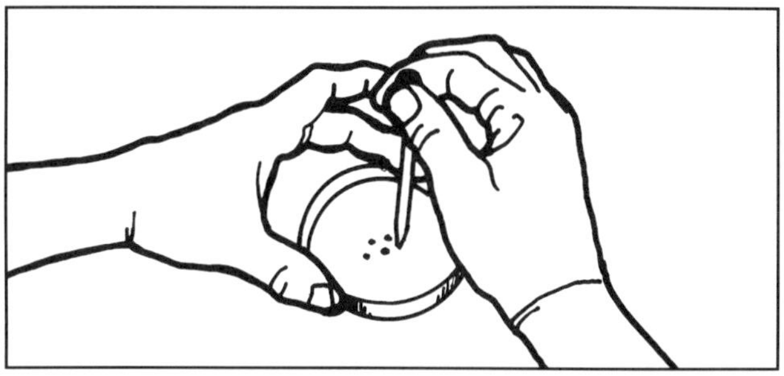

FIGURE 8-3b.

Boring Wells In the Saline Agar Plate *When cutting the wells in the agar, press straight down with the cutter; do not twist . Twisting the cutter or dropper will create fissures in the agar which could disrupt the diffusion of the sera.*

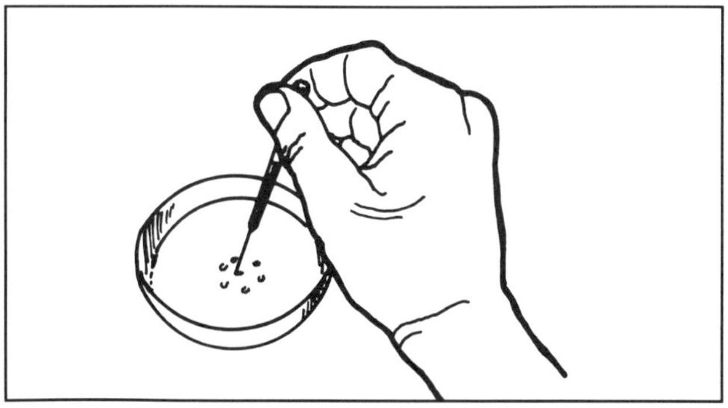

FIGURE 8-3c.

Filling the Wells *Place the tip of the pipette in the bottom of the well. Fill slowly to prevent creating air bubbles and to minimize spilling serum over the sides.*

5. Using a *different* micropipette for each transfer, carefully fill the wells as follows:
   a. Fill the center well with the anti-horse/anti-bovine mixture.
   b. Fill wells #1 and #2 with physiological saline.
   c. Fill wells #3 and #4 with horse serum.
   d. Fill wells #5 and #6 with bovine serum.
6. Cover the plate and allow it to sit undisturbed for 30 minutes.
7. Incubate at room temperature for up to 72 hours or until precipitation lines appear.
8. Examine the plate for precipitation patterns (Figure 8-3d).

Precautions

⚠ Be careful not to raise the agar from the plate as you remove the agar discs.

⚠ Overfilling the wells will spoil the results so be careful when adding the solutions.

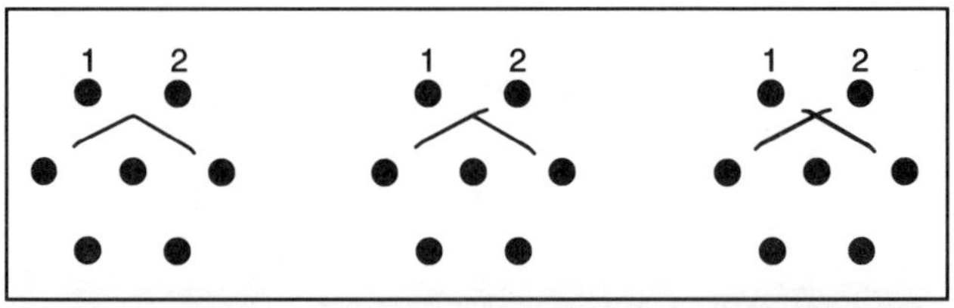

FIGURE 8-3d.

Precipitation Patterns *The diagram above shows three possible precipitation patterns formed between the antibody in the center well and the antigens in wells #1 and #2 . The pattern on the left demonstrates identity between the antigens. This indicates that the antigens are identical. The pattern in the middle demonstrates partial identity which means that the antigens are related but not identical. The pattern on the right shows nonidentity; the antigens are not related. Refer to Figure 8-21 in the* Photographic Atlas.

References

Lam, Joseph S. and Lucy M. Mutharia. 1994. Page 120 in *Methods for General and Molecular Bacteriology*, edited by Philipp Gerhardt, R. G. E. Murray, Willis A. Wood and Noel R. Krieg, American Society for Microbiology, Washington, DC.

Ouchterlony, O. 1968. Page 20 in *Handbook of Immunodiffusion and Immunoelectrophoresis*. Ann Arbor Science Publishers, Ann Arbor, MI.

Agglutination Reactions — Slide Agglutination

Photographic Atlas Reference

Page 104

Test Protocol

1. Using a grease pencil, draw two circles approximately the size of a dime on a microscope slide.
2. Place a drop of *Salmonella* anti-H antiserum in each circle.
3. Place a drop of Salmonella O antigen in one circle and a drop of Salmonella H antigen in the other circle. Using a *different* toothpick for each circle, mix until each of the antigens is completely emulsified with the antiserum. Discard the toothpicks in a biohazard container.
4. Allow the slide to sit for a few minutes and observe for agglutination.

Materials

Salmonella H antigen
Salmonella O antigen
Salmonella anti-H antiserum
One microscope slide
Toothpicks
Grease pencil

Precaution

⚠ Use a clean toothpick for each antigen to reduce the chance of false positives.

References

Collins, C. H., Patricia M. Lyne, J. M. Grange. 1995. Page 118 in *Collins and Lyne's Microbiological Methods, 7th Ed.* Butterworth-Heinemann, UK.

Lam, Joseph S. and Lucy M. Mutharia. 1994. Page 120 in *Methods for General and Molecular Bacteriology*, edited by Philipp Gerhardt, R. G. E. Murray, Willis A. Wood and Noel R. Krieg, American Society for Microbiology, Washington, DC.

Agglutination Reactions — Blood Typing

Photographic Atlas Reference

Page 104

Test Protocol

1. Place a drop of anti-A antiserum on one end of a microscope slide.
2. Place a drop of anti-B antiserum on the other end of the slide. Label the slide appropriately.
3. On the second microscope slide, place a drop of anti-Rh antiserum. Label the slide.
4. Clean the tip of your finger with an alcohol wipe. Let the alcohol dry.
5. Open a lancet package and remove the lancet, being careful not to touch the tip before you use it.
6. Prick the end of your finger and immediately place a drop of blood beside each drop of antiserum. Do not touch the antisera with your finger. It's OK to have someone else prick your finger, but make sure he or she wears protective gloves.
7. Discard the lancet in the sharps container.
8. Using a circular motion, mix each set of drops with a toothpick. Be sure to use a *different toothpick* for each antiserum.
9. Gently rock the slides back and forth for a few minutes or until agglutination occurs.
10. After the agglutination reaction is complete, record the results in the table below.
11. Using the information on pages 104 and 105 of the *Photographic Atlas,* determine your blood type and enter it in the space provided below.

Materials

Blood typing anti-A antiserum
Blood typing anti-B antiserum
Blood typing anti-Rh antiserum
Two microscope slides
Toothpicks
Grease pencil
Sterile lancets
Alcohol wipes
Small adhesive bandages
Sharps container
Disposable latex gloves

Precautions

⚠ Do not touch the antisera with your finger when placing the blood on the slide.

⚠ Although not absolutely essential, a "light box" used to gently warm the slides while you tilt them improves the hemagglutination reaction.

ANTISERUM	AGGLUTINATION +/−
Anti-A	
Anti-B	
Anti-Rh	

Blood Type ____________________

Enzyme Linked Immunosorbent Assay (ELISA)

Photographic Atlas Reference

Page 106

This test, a form of direct ELISA, is typically used by blood banks to screen donors' blood for the presence of Hepatitis B Surface Antigen. It differs from the examples included in the *Photographic Atlas* in that it makes use of antibody coated beads as the reaction medium. Refer to Figures 8-28 and 8-30 in the *Atlas* for further comparison.

Materials

Abbott Laboratories' Diagnostic Kit #1980-24 (AUSZYME® Monoclonal)

Materials Provided in the Kit

- Beads coated with Anti-HBs (antibody to Hepatitis B surface antigen)
- One vial Anti-HBs: Horseradish Peroxidase conjugate (antibody: enzyme conjugate)
- One vial Human HBsAg in TRIS buffer (positive control)
- One vial recalcified human plasma, nonreactive for Anti-HBs and HBsAg (negative control)
- One bottle OPD (*o*-Phenylenediamine • 2 HCl) tablets (substrate)
- One bottle diluent for OPD (citrate-phosphate buffer containing 0.02% hydrogen peroxide)
- Stopping reagent (1*N* sulfuric acid)
- Reaction trays
- Assay tubes with identifying racks
- Cover seals

Materials Required But Not Provided in the Kit

- Pipettes capable of delivering 50 μL, 200 μL, 300 μL and 1.0 mL
- Disposable graduated pipettes for measuring OPD diluent
- Dispensing pump and assembly for rinsing beads (Abbott Laboratories recommends *Abbott Quick Wash*)
- Water bath capable of maintaining a set point between 38 and 41°C
- Nonmetallic forceps
- Spectrophotometer capable of reading at 492 nm wavelength
- Unknown sample (optional)

Test Protocol

We have modified the manufacturer's recommended procedure to make it suitable for classroom use. **For clinical applications, please refer to the package insert.**

CAUTION! The positive and negative controls provided in this test kit are made from human blood components. Although they have been screened for infectious diseases they do not come with an absolute guarantee of safety. Handle the controls, and any specimens or unknowns provided by your instructor, as you would handle any potentially infectious substance. (See Precautions below.)

1. Allow all reagents and solutions to warm to room temperature before beginning the test procedure.
2. Prepare and maintain a data sheet throughout the experiment indicating the location and contents of each well including all appropriate times.
3. Gently swirl each reagent immediately before using to make sure that it is thoroughly mixed.
4. Dispense 200 μL volumes of the recalcified human plasma (negative control) into the bottoms of three wells in the reaction tray.
5. Dispense 200 μL volumes of the human HBsAg in TRIS buffer (positive control) into the bottoms of two wells in the reaction tray.
6. Dispense 200 μL volumes of any unknown samples into the tray in the same manner.
7. Add 50 μL of the Anti-HBs:Horseradish Peroxidase conjugate to each well containing a control and any wells containing unknown samples. Gently tap the tray to mix well.
8. Carefully add one bead to each well using the nonmetal forceps.
9. Apply the cover seal. Gently tap the tray to dislodge any trapped air bubbles.
10. Incubate the tray according to one of the following procedures: *

PROCEDURE	TEMPERATURE	TIME
A	38 to 41°C	3 hours
B	15 to 30°C	12 to 20 hours

*The two procedures differ only in the amount of time and the temperature for incubation. If you are unsure which procedure to follow, ask your instructor then continue with the test.

11. Immediately before completion of the incubation period, prepare the OPD substrate solution as follows:
    a. Using clean pipettes and containers, add one tablet for every 5.0 mL of the diluent used. (Five milliliters of OPD solution is enough for approximately 13 tests.)
    b. Allow the tablet to dissolve in the diluent. (Because the OPD solution quickly deteriorates, you must complete steps 11 through 15 in 60 minutes or less.)
12. Remove the tray from the incubator and discard the cover seal. Aspirate the liquid and wash each bead three to five times with 4 to 6 mL distilled or deionized water. Total rinse volume for each bead should be 12 to 30 mL.
13. Immediately transfer beads to appropriately labeled assay tubes. To do this, invert the rack containing the appropriately positioned assay tubes, press the tubes tightly against the reaction tray and invert the tray and tubes together. Gently tap the tray until each of the beads has fallen into its corresponding assay tube.
14. Gently swirl the OPD substrate solution until the tablet is thoroughly mixed.
15. Immediately pipette 300 μL of the freshly prepared substrate solution into two empty assay tubes (to be used as blanks) and then into each tube containing a bead.
16. Cover the tubes and incubate at room temperature for 30 minutes. This is the *color development* period.
17. Add 1.0 mL of 1*N* Sulfuric acid stopping reagent to each tube. This will stop the enzymatic reaction and prepare the assay tubes for reading in the spectrophotometer.
18. Blank the spectrophotometer (with the wavelength set at 492 nm) using one of the tubes containing only OPD substrate solution and sulfuric acid. If you need help blanking the spectrophotometer, ask your instructor for assistance.
19. Immediately measure the absorbance of each tube containing a bead and record your results below.

SAMPLE	ABSORBANCE READING
Negative Control #1	
Negative Control #2	
Negative Control #3	
Positive Control #1	
Positive Control #2	
Unknown #1	
Unknown #2	

Calculations and Data Analysis

You can determine the presence or absence of HBsAg in an unknown sample by comparing its absorbance value to the *cutoff value* (described in #2 below). The cutoff value (an absorbance value) is the point below which results can confidently be interpreted as negative for HBsAg and above which results can confidently be interpreted as positive for HBsAg. The cutoff value is the sum of 0.050, which is a value derived from actual blood bank data, and the average of the negative control values called the *negative control mean* (NCm).

The NCm (described in #1 below) is the three negative control values added together divided by 3. This value is used as a baseline for comparison of real specimens and is also used in conjunction with the *positive control mean* (PCm) to determine test accuracy.

The PCm (described in #3 below) is the two positive control values added together divided by 2. Subtracting the NCm from the PCm produces what is called the *P-N value.*

The positive controls are manufactured to produce absorbance readings at least 0.400 higher than the negative controls. Therefore, the P-N value (described in #4 below) must be equal to or greater than 0.400 for the test run to be valid. If the P-N value is below 0.400, testing technique is the likely cause, and the test must be repeated. If the P-N value is consistently below 0.400, reagent deterioration may be the cause.

1. Calculating the NCm

To calculate the NCm add the three negative control absorbance readings and divide the total by 3.
For example:

NEGATIVE CONTROL SAMPLE #	ABSORBANCE
1	0.010
2	0.011
3	0.009
TOTAL	0.030

$$\text{NCm} = \frac{0.030}{3}$$

$$\text{NCm} = 0.010$$

Note Negative control values should be equal to or less than 0.100 and equal to or more than –0.010. Also, negative control values should be equal to or greater than 0.5 times the NCm and equal to or less than 1.5 times the NCm. If one value does not fall within this range, it should be discarded and the mean recalculated. If two values do not fall within this range, the test should be done over.

2. Calculating the cutoff value

Add 0.050 to the negative control mean which, from the sample calculation above, is 0.010. Samples which are equal to or above this value are considered to be positive for hepatitis B surface antigen (HBsAg). Therefore:

$$\text{Cutoff Value} = \text{NCm} + 0.050$$
$$\text{Cutoff Value} = 0.010 + 0.050$$
$$\text{Cutoff Value} = 0.060$$

In this example, any specimen with an absorbance reading equal to or greater than 0.060 would be suspected reactive (+) for HBsAg. It would then be subjected to repeat testing in addition to a confirmatory assay before final determination would be made.

3. Calculating the positive control mean (PCm)

Add the two positive control absorbance readings and divide the total by 2.

For example:

POSITIVE CONTROL SAMPLE #	ABSORBANCE
1	1.024
2	1.030
TOTAL	2.054

$$\text{PCm} = \frac{2.054}{2}$$

$$\text{PCm} = 1.027$$

As will be demonstrated in #4 below, the positive control mean is necessary for assuring the validity of the test itself.

4. Calculating the P-N value

Deteriorating chemicals and/or human error can produce inaccurate results. For this reason you must calculate the P-N value before determining the reactivity or nonreactivity of a sample. Do this by subtracting the NCm from the PCm. As stated above, the P-N value must be equal to or greater than 0.400 for the test run to be valid. Using the figures from the above examples (where the NCm is 0.010 and the PCm is 1.027) calculate the P-N as follows:

$$\text{P} - \text{N Value} = \text{PCm} - \text{NCm}$$
$$\text{P} - \text{N Value} = 1.027 - 0.010$$
$$\text{P} - \text{N Value} = 1.017$$

In this example the P-N value is greater than 0.400. Therefore, the test run is valid.

Precautions

- ⚠ Make certain that all reagents are at room temperature before beginning the test.
- ⚠ Gently swirl each reagent immediately before using to make sure that it is thoroughly mixed.
- ⚠ To reduce the possibility of contamination, use absolutely clean pipettes and containers.
- ⚠ Always wear disposable gloves while working with these materials and wash your hands when you are finished.
- ⚠ Work carefully so as not to splash any of the reagents or acid onto the work surface or yourself.
- ⚠ Time is critical in this test. Organize the materials and equipment so that the test can be conducted without interruption and that all time limits can be adhered to.
- ⚠ The OPD solution should be pale yellow or colorless. A yellow-orange solution indicates possible contamination and should be discarded.
- ⚠ Avoid strong light during the color development period.
- ⚠ Dispense the acid in the same sequence as you dispensed the OPD solution. Since it is impossible to add solution to all the tubes at the same time, this procedure helps to compensate for the time elapsed while working on several tubes.
- ⚠ Do not allow any of the chemicals or reagents in this kit to come in contact with metal.
- ⚠ Dispose of all pipettes and containers in the appropriate autoclave or biohazard receptacle.
- ⚠ A more extensive list of precautions appropriate to the clinical use of this product can be found on the package insert.

Reference

Package insert, Abbott Laboratories Test Kit # 1980-24 (AUSZYME® Monoclonal). Published in USA February 1990.

SECTION NINE

Viral, Protozoan and Fungal Microbiology

T4 Virus and HIV

Photographic Atlas Reference

Pages 109

Procedure

Viral structure is not typically studied in the laboratory of an introductory microbiology course. Refer to your textbook (see References) for electron micrographs to study viral structure or to the plaque assay laboratory as an example of a technique using viruses.

Materials

Textbook

References

Alcamo, I. Edward. 1994. Chapter 11 in *Fundamentals of Microbiology, 4th Ed.* Benjamin/Cummings Publishing Company, Inc., Redwood City, CA.

Atlas, Ronald M. 1995. Chapter 8 in *Principles of Microbiology.* Mosby-Year Book, Inc., St. Louis, MO.

Black, Jaqueline G. 1996. Chapter 11 in *Microbiology Principles and Applications, 3rd Ed.* Prentice-Hall, Inc. Upper Saddle River, NJ.

Ingraham, John L. and Catherine A. Ingraham. 1995. Chapter 13 in *Introduction to Microbiology.* Wadsworth Publishing Company, Belmont, CA.

McKane, Larry and Judy Kandel. 1996. Chapter 13 in *Microbiology Essentials and Applications, 2nd Ed.* McGraw-Hill, Inc., New York, NY.

Nester, Eugene W., C. Evans Roberts and Martha T. Nester. 1995. Chapters 13 and 14 in Microbiology A Human Perspective. Wm. C. Brown Publishers, Dubuque, IA.

Prescott, Lansing M., John P. Harley, and Donald A. Klein. 1996. Chapters 16, 17 and 18 to 24 hour pure cultures of: in *Microbiology, 3rd Ed.* Wm. C. Brown Publishers, Dubuque, IA.

Talaro, Kathleen and Arthur Talaro. 1993. Chapter 5 in *Foundations in Microbiology.* Wm. C. Brown Publishers, Dubuque, IA.

Tortora, Gerard J., Berdell R. Funke and Christine L. Case. 1995. Chapter 13 in *Microbiology — An Introduction, 5th Ed.* Benjamin Cummings Publishing Company, Inc., Redwood City, CA.

Protozoans

Photographic Atlas Reference

Page 112

Procedure

1. Make wet mount preparations of the living specimens as illustrated in Figures 9-1a and 9-1b. You may wish to add a drop of methylene blue to stain the organisms. A drop of methyl cellulose may also be useful as this slows down the fast swimmers. Observe the following structures on each organism. Sketch and label these in the table below.

   Amoeba: nucleus, pseudopods, vacuoles

   Paramecium: macronucleus, micronucleus, cilia, oral groove, contractile vacuole

   Euglena: nucleus, flagellum

Materials

Fresh culture of *Amoeba spp.*
Fresh culture of *Paramecium spp.*
Fresh culture of *Euglena spp.*
Pond water
Methyl cellulose
Clean microscope slides and cover glasses
Methylene blue stain
Prepared slides of:

- *Entamoeba histolytica* trophozoite and cyst
- *Balantidium coli* trophozoite and cyst
- *Giardia lamblia* trophozoite and cyst
- *Trichomonas vaginalis* trophozoite
- *Leishmania donovani* promastigote
- *Trypanosoma spp.*
- *Plasmodium spp.*
- *Toxoplasma gondii* trophozoite

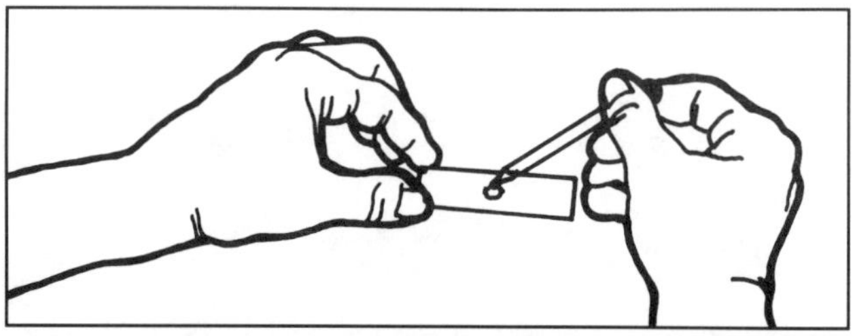

FIGURE 9-1a.

Place a Drop of Water on the Slide *Use your loop or a dropper to place a small amount of water containing the specimens on a clean slide.*

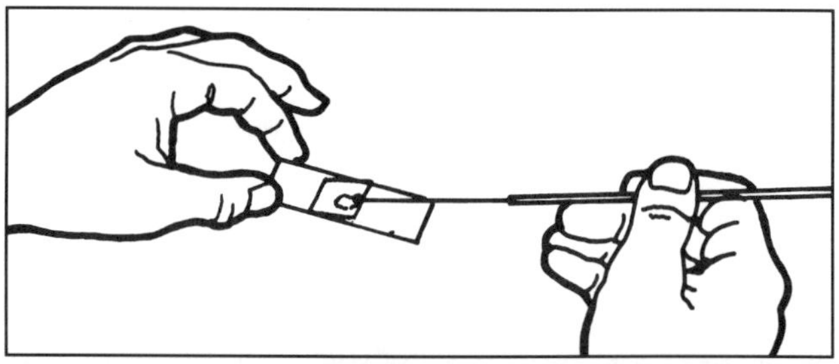

FIGURE 9-1b.

Lower the Cover Glass *Use your loop to gently lower the cover glass onto the water drop. Avoid trapping air bubbles under the cover glass.*

ORGANISM	MAGNIFICATION	LABELED SKETCH
Amoeba		
Paramecium		
Euglena		

2. Obtain prepared slides of the protozoan pathogens and observe them under appropriate magnification. You should observe the following structures on each organism. (Many of these slides are made from patient samples, so there will be a lot of other material on the slide besides the desired organism. You must search carefully and with patience.) Sketch and label these in the table below.

   Entamoeba histolytica trophozoite: pseudopods, nucleus with karyosome, ingested erythrocytes

   Entamoeba histolytica cysts: multiple nuclei (up to four) with karyosomes

   Balantidium coli trophozoite: cilia, macronucleus, micronucleus

   Balantidium coli cyst: multiple nuclei

   Giardia lamblia trophozoite: flagella (four pairs), sucking disc, nuclei (two), median bodies (two)

   Giardia lamblia cyst: multiple nuclei (four), median bodies (four)

   Trichomonas vaginalis trophozoite: nucleus, flagella (four)

   Leishmania donovani promastigote: flagellum, nucleus

   Trypanosoma spp.: nucleus, flagellum, undulating membrane

   Plasmodium spp.: ring stage, mature trophozoite, schizont, male gametocyte, female gametocyte

   Toxoplasma gondii trophozoite (tachyzoite): nucleus

References

Garcia, Lynne S., Alexander J. Sulzer, George Healy, Katharine K. Grady, and David A. Bruckner. 1995. Chapter 104 in *Manual of Clinical Microbiology, 6th Ed.*, edited by Patrick R. Murray, Ellen Jo Baron, Michael A. Pfaller, Fred C. Tenover, and Robert H. Yolken. American Society for Microbiology, Washington, D.C.

Healy, George R. and Lynne S. Garcia. 1995. Chapter 106 in *Manual of Clinical Microbiology, 6th Ed.*, edited by Patrick R. Murray, Ellen Jo Baron, Michael A. Pfaller, Fred C. Tenover, and Robert H. Yolken. American Society for Microbiology, Washington, D.C.

Lee, John J., Seymour H. Hutner, and Eugene C. Bovee. 1985. *Illustrated Guide to the Protozoa.* Society of Protozoologists, Lawrence, KS.

Markell, Edward K., Marietta Voge and David T. John. 1992. *Medical Parasitology, 7th Ed.* W.B. Saunders Company, Philadelphia, PA.

ORGANISM	MAGNIFICATION	LABELED SKETCH
Entamoeba histolytica trophozoite and cyst		
Balantidium coli trophozoite and cyst		
Giardia lamblia trophozoite and cyst		
Trichomonas vaginalis trophozoite		
Leishmania donovani promastigote		
Trypanosoma spp.		
Plasmodium spp.		
Toxoplasma gondii trophozoite		

Fungi

Photographic Atlas Reference

Page 120

Materials

Agar slant of *Saccharomyces cerevisiae*
Plate culture of *Aspergillus spp.*
Plate culture of *Penicillium spp.*
Plate culture of *Rhizopus spp.*
Dissecting microscope
Prepared slides of:

- *Aspergillus spp.* conidiophore
- *Candida albicans*
- *Penicillium spp* colony
- *Penicillium spp.* conidiophore
- *Pneumocystis carinii*
- *Rhizopus spp.* sporangia
- *Rhizopus spp.* gametangia
- *Saccharomyces cerevisiae* ascospores

Procedure

Yeasts

1. Make a wet mount slide of *Saccharomyces cerevisiae* and observe under high dry and oil immersion. Identify vegetative cells and budding cells. Sketch what you see in the table provided.
2. Observe prepared slides of *Saccharomyces cerevisiae* ascospores. Sketch what you see in the table provided.
3. Observe prepared slides of *Candida albicans*. Identify vegetative cells and budding cells. Sketch what you see in the table below.

Molds

1. Observe the plate culture of *Rhizopus* using the dissecting microscope. Identify hyphae, rhizoids, and sporangia. Sketch and label what you see in the table below.

ORGANISM	MAGNIFICATION	LABELED SKETCH
S. cerevisiae vegetative cells and ascospores		
Candida albicans		

ORGANISM	MAGNIFICATION	LABELED SKETCH
Rhizopus culture		
Rhizopus sporangia		
Rhizopus gametangia		

2. Examine prepared slides of *Rhizopus* sporangia using medium and high dry powers. Identify the following: sporangiophores, sporangia, and spores. Sketch and label what you see in the table provided.
3. Examine prepared slides of *Rhizopus* gametangia using medium and high dry power. Identify the following: progametangia, gametangia, young zygosporangia, mature zygosporangia. Sketch and label what you see in the table provided.
4. Observe the plate culture of *Penicillium* using the dissecting microscope. Identify hyphae and ascospores. Sketch and label what you see in the table provided.
5. Observe prepared slides of *Penicillium*. Identify the following: hyphae, asci, and ascospores. Sketch and label what you see in the table provided.
6. Observe prepared slides of *Penicillium* conidiophores. Identify the following: hyphae, conidiophores and conidia.
7. Observe the plate culture of *Aspergillus* using the dissecting microscope. Identify hyphae and ascospores. Sketch and label what you see in the table provided.
8. Observe prepared slides of *Aspergillus* conidiophores. Identify hyphae, conidiophores and conidia. Sketch and label what you see in the table provided.
9. Observe prepared slides of *Pneumocystis carinii* and identify the following: cysts, intracystic bodies (up to eight), and trophic forms. Sketch and label what you see in the table provided.

ORGANISM	MAGNIFICATION	LABELED SKETCH
Penicillium culture		
Penicillium ascus		
Penicillium conidiophore		

ORGANISM	MAGNIFICATION	LABELED SKETCH
Aspergillus culture		
Aspergillus conidiophore		

ORGANISM	MAGNIFICATION	LABELED SKETCH
Pneumocystis cysts and trophic forms		

References

Collins, C.H., Patricia M. Lyne and J.M. Grange. 1995. Chapter 51 in *Collins and Lyne's Microbiological Methods, 7th Ed.* Butterworth-Heineman, Oxford.

Hadley, W. Keith and Valerie L. Ng. 1995. Chapter 62 in *Manual of Clinical Microbiology, 6th Ed.*, edited by Patrick R. Murray, Ellen Jo Baron, Michael A. Pfaller, Fred C. Tenover, and Robert H. Yolken. American Society for Microbiology, Washington, D.C.

Kennedy, Michael J. and Lynne Sigler. 1995. Chapter 64 in *Manual of Clinical Microbiology, 6th Ed.*, edited by Patrick R. Murray, Ellen Jo Baron, Michael A. Pfaller, Fred C. Tenover, and Robert H. Yolken. American Society for Microbiology, Washington, D.C.

Mauseth, James D. 1995. Chapter 20 in Botany An Introduction to Plant Biology, 2nd Ed. Saunders College Publishing, Philadelphia, PA.

Raven, Peter H., Ray F. Evert and Susan Eichhorn. 1992. Chapter 12 in *Biology of Plants, 5th Ed.* Worth Publishers, New York, NY.

Warren, Nancy G. and Kevin C. Hazen. 1995. Chapter 61 in *Manual of Clinical Microbiology, 6th Ed.*, edited by Patrick R. Murray, Ellen Jo Baron, Michael A. Pfaller, Fred C. Tenover, and Robert H. Yolken. American Society for Microbiology, Washington, D.C.

APPENDIX A

Basic Medium Recipes

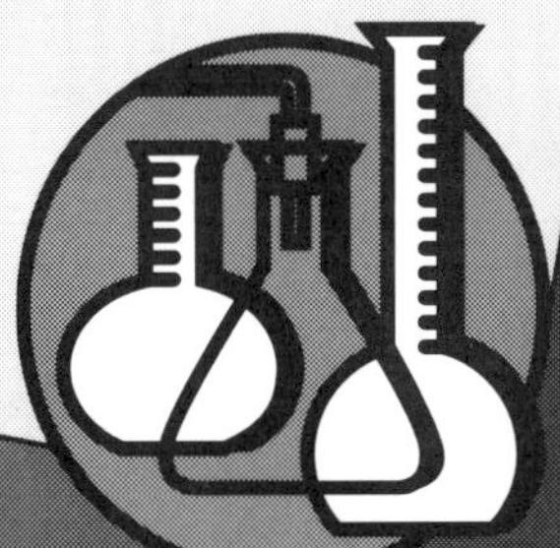

Nutrient Agar and Nutrient Broth

Recipes

Nutrient Agar

Beef extract	3.0 g
Peptone	5.0 g
Agar	15.0 g
Distilled or deionized water	1.0 L
final pH = 6.8 ± 0.2 at 25°C	

Nutrient Broth

Beef extract	3.0 g
Peptone	5.0 g
Distilled or deionized water	1.0 L
final pH = 6.8 ± 0.2 at 25°C	

Media Preparation

Nutrient Agar Tubes

1. Suspend the ingredients in one liter of distilled or deionized water and mix well.
2. Dissolve the mixture by heating with agitation.
3. The ingredients should be completely dissolved by boiling the mixture for 1 minute.
4. Dispense 7.0 mL portions (for slants) or 10 mL portions (for stabs) into test tubes and cap loosely.
5. Autoclave for 15 minutes at 121°C to sterilize the medium.
6. Cool to room temperature with the tubes in an upright position for agar deep tubes. Cool with the tubes on an angle for agar slant tubes.

Nutrient Agar Plates

1. Suspend the ingredients in one liter of distilled or deionized water and mix well.
2. Dissolve the mixture by heating with agitation.
3. The ingredients should be completely dissolved by boiling the mixture for 1 minute.
4. Autoclave for 15 minutes at 121°C to sterilize the medium.
5. Allow the agar to cool to 50°C.
6. Dispense approximately 15 mL into sterile Petri plates. Gently swirl to get the agar to completely cover the base.
7. Allow to cool before inoculating.

Nutrient Broth

1. Suspend the ingredients in one liter of distilled or deionized water. Mix to dissolve.
2. Continue dissolving the mixture by heating while agitating it, if necessary.
3. Dispense 7.0 mL portions into test tubes.
4. Autoclave for 15 minutes at 121°C to sterilize the medium.

Precautions

- ⚠ To minimize contamination, clean the work surface, turn off all fans and close any doors that might allow excessive air movements.
- ⚠ Shield the Petri dish with its lid while you pour agar to reduce the chance of introducing airborne contaminants.

References

DIFCO Laboratories. 1984. Pages 619 and 622 in *DIFCO Manual, 10th Ed.* DIFCO Laboratories, Detroit, MI.

Power, David A. and Peggy J. McCuen. 1988. Pages 214 and 215 in *Manual of BBL® Products and Laboratory Procedures, 6th Ed.* Becton Dickinson Microbiology Systems, Cockeysville, MD.

APPENDIX B

Microbial Transfer Methods

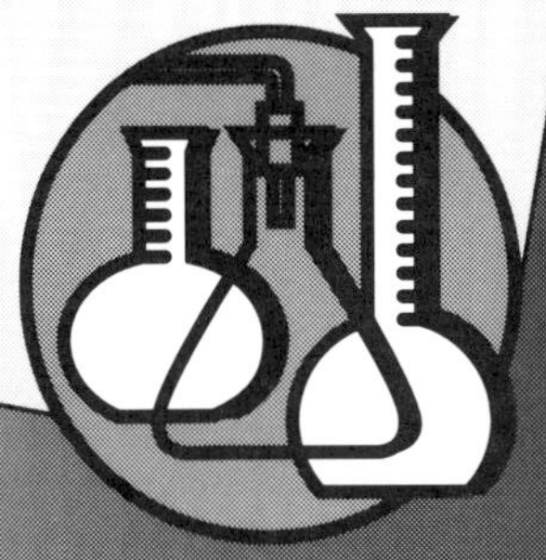

As a microbiologist, you may be required to collect a specimen and then transfer it from its source to a sterile medium *aseptically* (without contamination by other microbes). In the laboratory, you may be required to transfer a pure culture of an organism to another sterile medium, again, without contamination. Good transfer technique also limits contamination of surroundings by the cultured organism. These transfers are basic to most work you will be doing this semester, so practice them until you become proficient.

To prevent contamination of the sample, inoculating instruments (see Figure B-1) are sterilized prior to use. Typically, they are either autoclaved (*e.g.*, cotton swabs, Pasteur pipettes, and serological or Mohr pipettes) or incinerated in a Bunsen burner flame (*e.g.*, inoculating loops or needles). The former are sterilized long before use, whereas the latter are sterilized immediately before use. Glass tubes or flasks containing a culture are autoclaved along with the medium they contain, but the opening is incinerated again at the time of transfer.

This section is divided into two main parts describing the two main stages of transfers. These are: 1) obtaining the sample to be transferred, and 2) transferring to the sterile culture medium. As shown in Figure B-2, you may transfer cells from most any medium to most any other medium. Although aseptic transfers are not difficult, preparation will make whichever one you use go much more smoothly. You need to know where the sample is coming from, the type of transfer instrument to be used, and the sample's destination before you begin. Until transfers become second nature to you, we recommend pulling the appropriate pages for obtaining and transferring the sample and keeping them in a visible place.

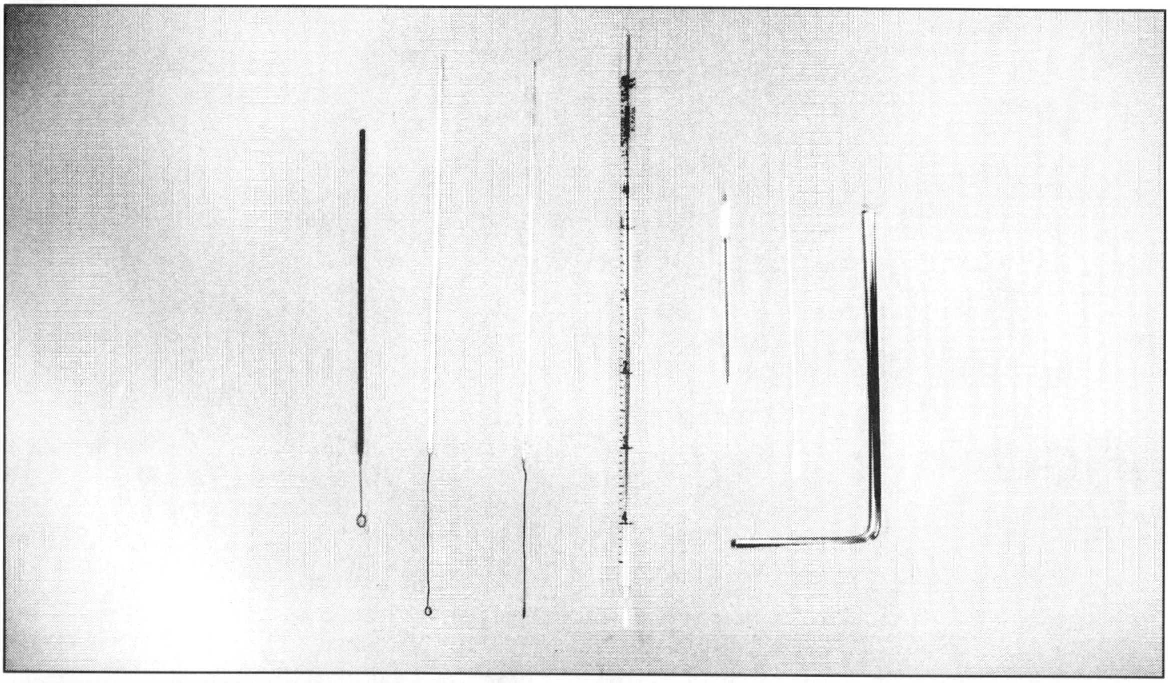

FIGURE B-1.

Inoculating Tools *Many different instruments may be used to transfer a microbial sample, the choice of which depends on the sample source, its destination, and any requirements imposed by specific procedures. Shown here are several examples of transfer instruments. From left to right: a disposable inoculating loop, nichrome inoculating loop, inoculating needle, serological pipette, Pasteur pipette, cotton swab and glass spreading rod.*

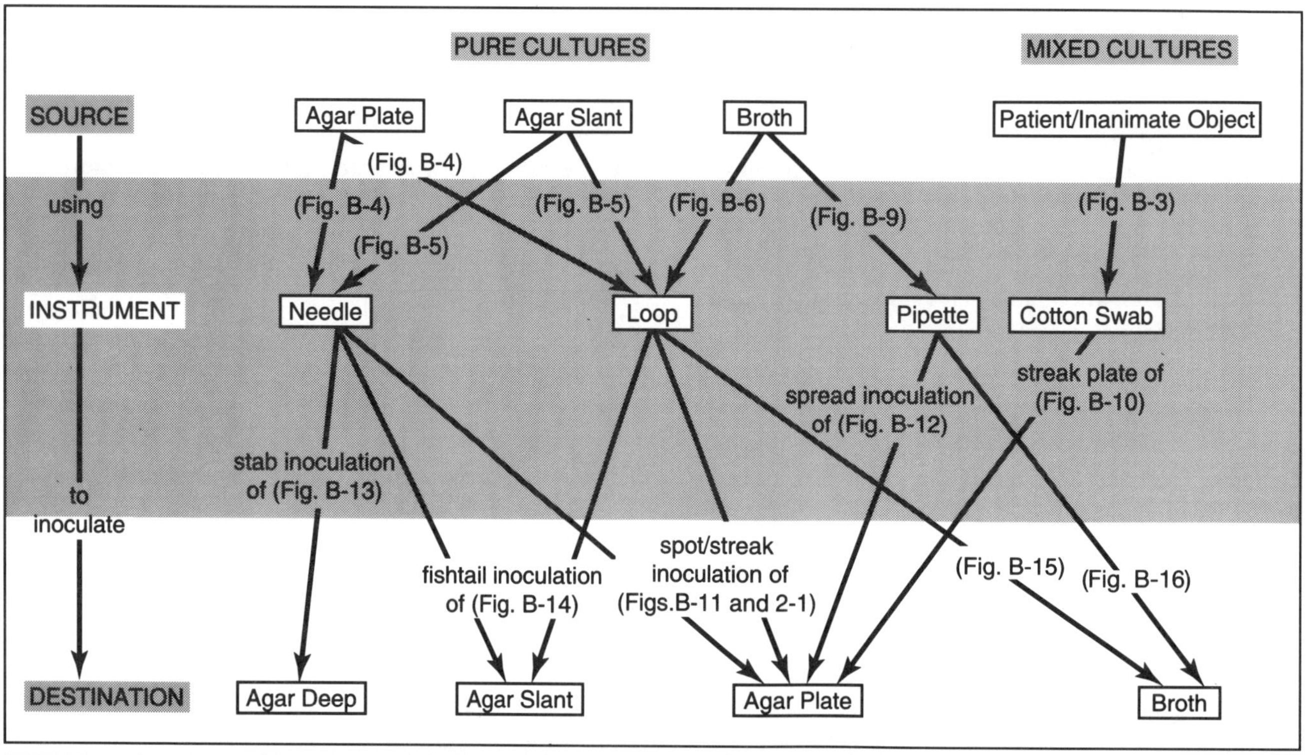

FIGURE B-2.

Typical Inoculations *Use this as a guide for choosing an appropriate transfer instrument for a given inoculation. Figure numbers provide the location of instructions on how to perform each transfer.*

Obtaining the Sample to be Transferred

Transfers with a Sterile Cotton Swab

A cotton swab is usually used to obtain a sample from a patient or an inanimate object, rather than from a culture grown in the laboratory. Sterile swabs may be dry or they may be in sterile water, depending on the sample source. In either case, care must be taken not to contaminate the swab by touching other surfaces with it. Examples are illustrated in Figures B-3a and B-3b, but your instructor may provide specific instructions on sample collection from other sources with a swab.

FIGURE B-3a.

Sampling a Patient's Throat *A tongue depressor and swab prepared in sterile water may be used together to obtain a sample from the pharynx (throat). Touching other parts of the oral cavity is likely to cause contamination. Also, avoid touching the soft palate or a gag reflex may be initiated! The sample should then be transferred from the swab to a growth medium as quickly as possible. Your instructor may provide instructions on sampling other body regions.*

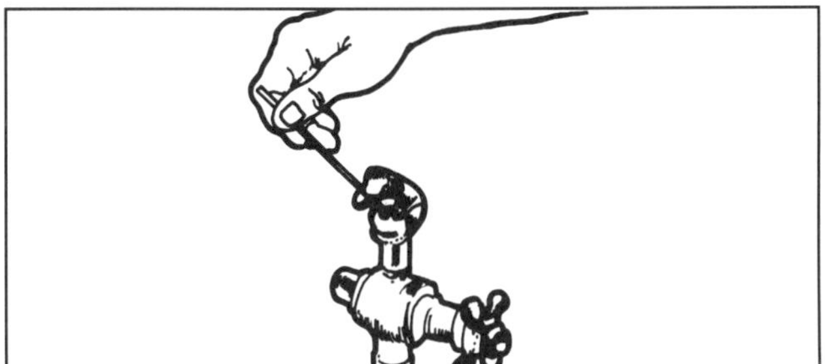

FIGURE B-3b.

Sampling an Inanimate Object *A swab prepared in sterile water may be used to obtain a sample from an environmental source. As with the throat culture, be careful to touch only the area to be sampled and transfer to the growth medium as soon as possible.*

Transfers From an Agar Plate Using an Inoculating Loop or Needle

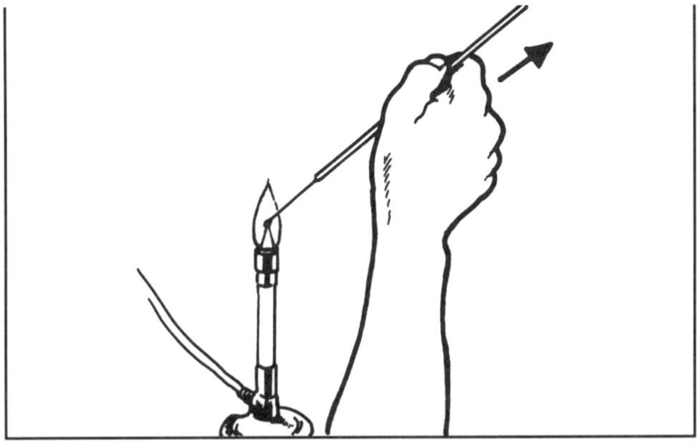

FIGURE B-4a.

Flame the Loop/Needle *Sterilize the loop/needle by incinerating it in the Bunsen burner flame. Hold the handle like a pencil in your dominant hand and relax! Pass it through the tip of the inner cone of flame (the hottest part) holding it in a nearly vertical position (approximately 70 to 80°). Begin flaming about 2 cm up the handle, then proceed down the wire to the tip. Flaming in this direction limits aerosol production by allowing the tip to heat up more slowly than if it were thrust into the flame immediately. Be sure to heat the entire wire to red-hot.*

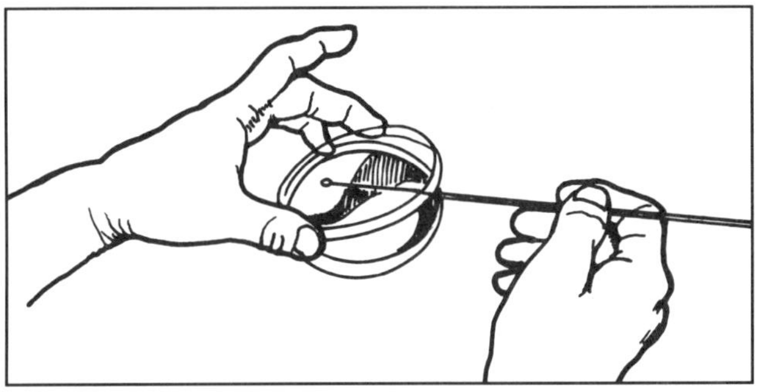

FIGURE B-4b.

Use the Lid as a Shield *Lift the lid of the agar plate, but still use it as a cover to prevent contamination from above. Touch the loop/needle to an uninoculated portion of the plate to cool it. (Placing a hot wire on growth may cause spattering of the growth and create aerosols.) Touch the wire to a colony of the desired organism and obtain a small amount of growth. Carefully remove the loop/needle and replace the lid, holding the loop/needle still. (Excessive movement of the loop/needle may cause aerosols.) What you do next depends on the medium to which you are transferring the growth. Please continue with the appropriate inoculation section.*

Transfers From an Agar Slant Using an Inoculating Loop or Needle

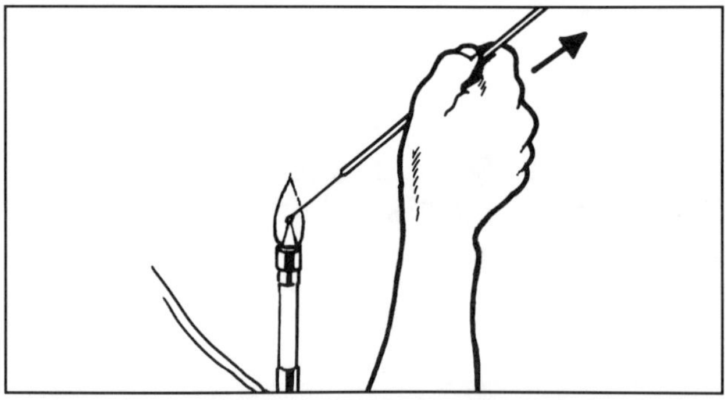

FIGURE B-5a.

Flame the Loop/Needle *Sterilize the loop/needle by incinerating it in the Bunsen burner flame. Hold the handle like a pencil in your dominant hand and relax! Pass it through the tip of the inner cone of flame (the hottest part) holding it in a nearly vertical position (approximately 70 to 80°). Begin flaming about 2 cm up the handle, then proceed down the wire to the tip. Flaming in this direction limits aerosol production by allowing the tip to heat up more slowly than if it were thrust into the flame immediately. Be sure to heat the entire wire to red-hot.*

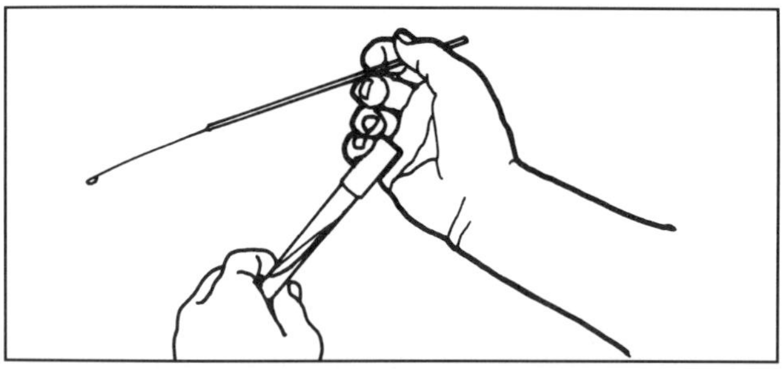

FIGURE B-5b.

Hold the Cap in Your Pinkie Finger *While keeping the loop hand still and away from the flame), bring the culture tube toward your other hand. Use your pinkie finger to remove and hold its cap. (The cap may be loosened prior to the transfer, especially if it's a screw top cap.)*

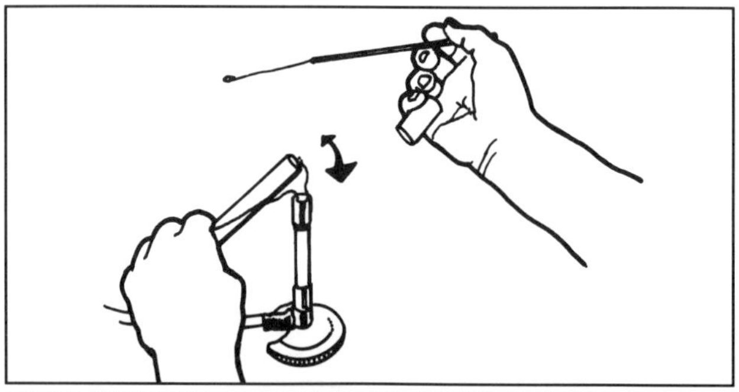

FIGURE B-5c.

Sterilize the Tube *Pass the lip of the tube quickly through the flame two or three times to sterilize the glass and the surrounding air. The tube should be held on a 45° angle to prevent contamination from above. Keep your loop/needle hand still.*

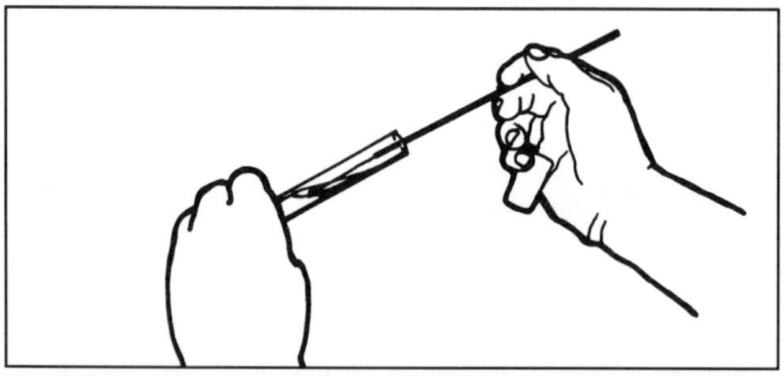

FIGURE B-5d.

Harvest the Growth *Hold the open tube with the agar surface up and at a 45° angle to prevent aerial contamination. Holding the loop/needle hand still, move the tube up the wire until the wire tip is over the desired growth. Touch the loop/needle to the growth and get the smallest visible mass of growth on the loop/needle. Then, holding the loop hand still,* move the tube *to remove the wire. Be especially careful when removing the tube not to catch the loop/needle tip on the tube lip. This springing action of the loop creates bacterial aerosols.*

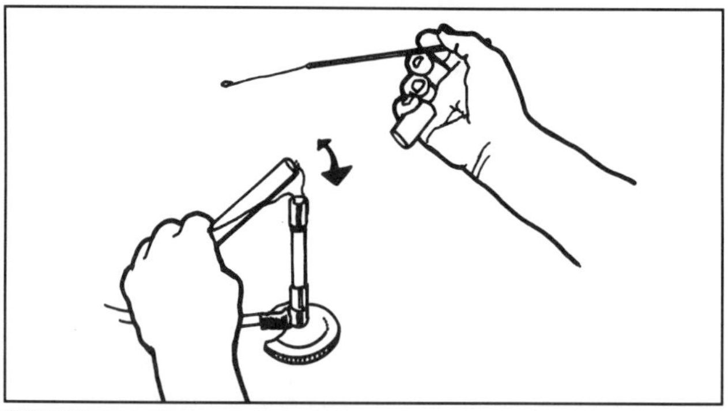

FIGURE B-5e.

Sterilize the Tube Again *Flame the tube lip as before. Keep your loop/needle hand still.*

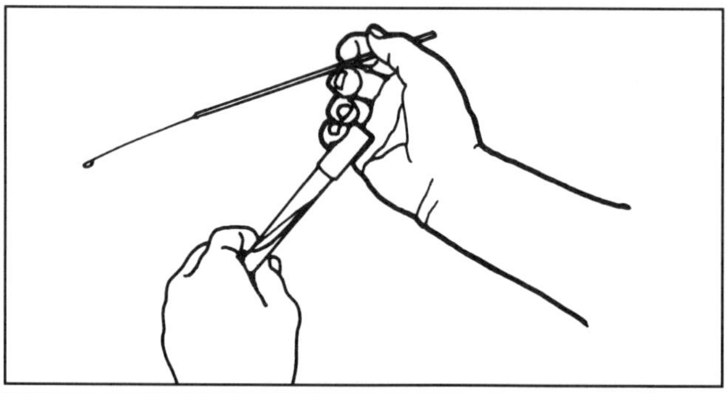

FIGURE B-5f.

Replace the Cap *Keeping the loop/needle hand still (remember, it has growth on it), move the tube to replace its cap. The cap at this point doesn't need to be on firmly — just enough to cover the tube. What you do next depends on the medium to which you are transferring the growth. Please continue with the appropriate inoculation section.*

Transfers From a Broth Culture Using a Loop

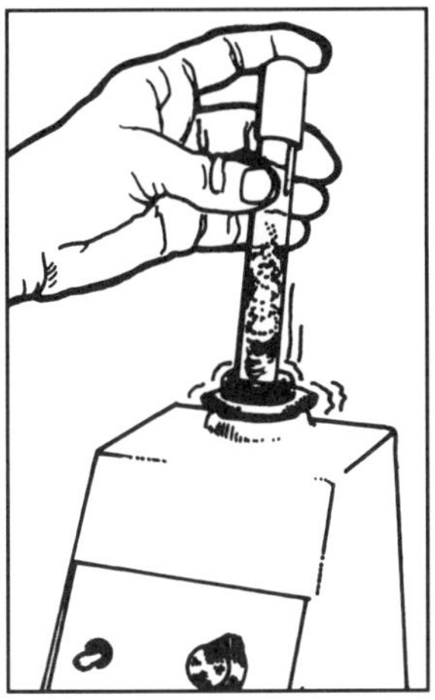

FIGURE B-6a.

Suspend the Bacteria in the Broth (One Method) *Growth may be suspended in the broth with a vortex mixer. Be sure not to mix so vigorously that broth gets into the cap or that you lose control of the tube. It's best to start slowly, then gently increase the speed until the tip of the vortex reaches the bottom of the tube.*

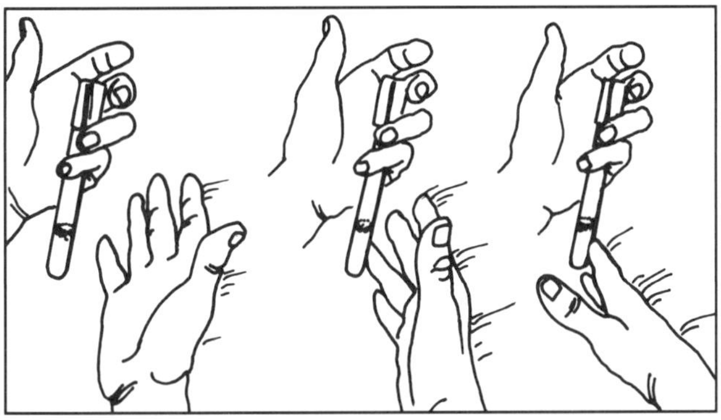

FIGURE B-6b.

Suspend the Bacteria in the Broth (Another Method) *The broth may also be agitated by drumming your fingers along the length of the tube several times. Be careful not to splash the broth into the cap or lose control of the tube.*

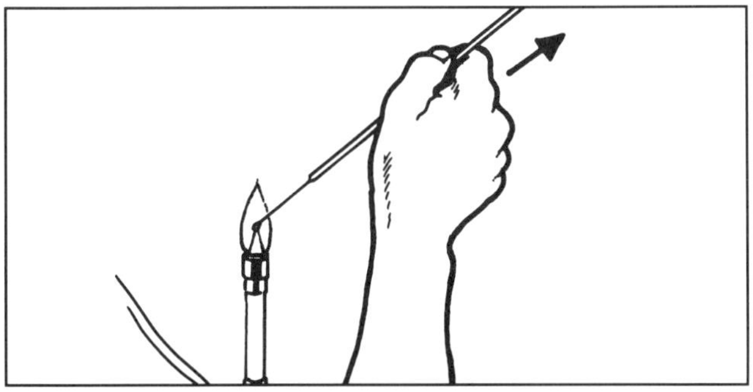

FIGURE B-6c.

Flame the Loop *Sterilize the loop by incinerating it in the Bunsen burner flame. Hold the handle like a pencil in your dominant hand and relax! Pass it through the tip of the inner cone of flame (the hottest part) holding it in a nearly vertical position (approximately 70 to 80°). Begin flaming about 2 cm up the handle, then proceed down the wire to the tip. Flaming in this direction limits aerosol production by allowing the tip to heat up more slowly than if it were thrust into the flame immediately. Be sure to heat the entire wire to red-hot.*

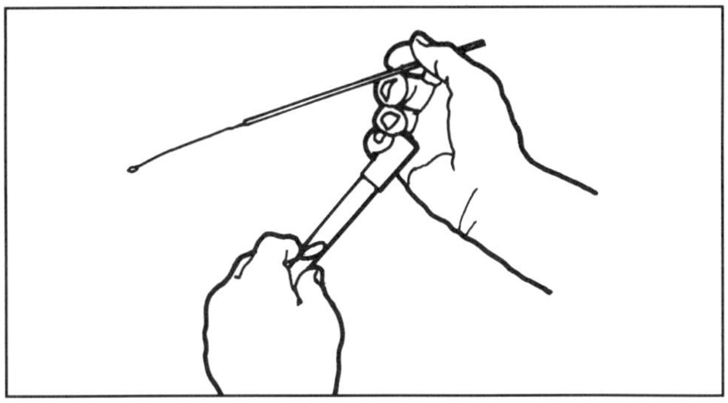

FIGURE B-6d.

Hold the Cap in Your Pinkie Finger *While keeping the loop hand still and away from the flame, bring the culture tube toward your other hand. Use your pinkie finger to remove and hold its cap. (The cap may be loosened prior to the transfer, especially if it's a screw top cap.)*

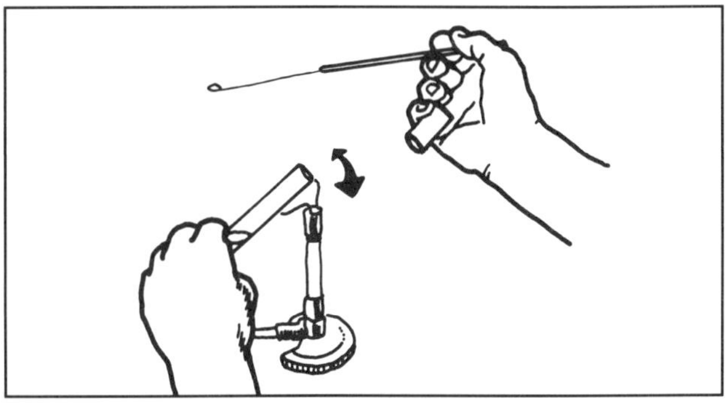

FIGURE B-6e.

Sterilize the Tube *Pass the lip of the tube quickly through the flame two or three times to sterilize the glass and the surrounding air. The tube should be held on a 45° angle to prevent contamination from above. Keep your loop hand still.*

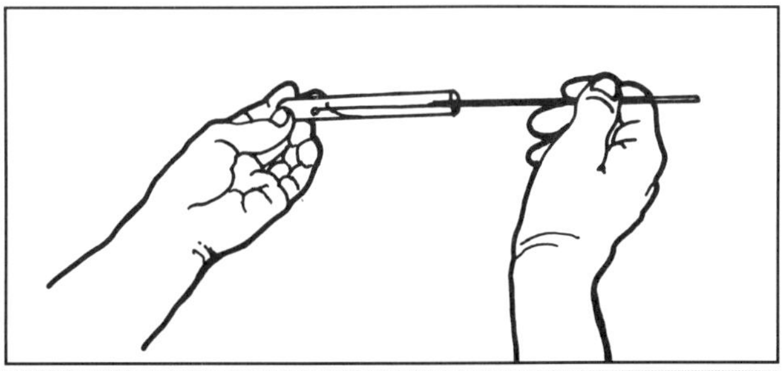

FIGURE B-6f.

Harvest the Growth *Hold the open tube at an angle to prevent aerial contamination. Holding the loop hand still, move the tube up the wire until the tip is in the broth. Then, holding the loop hand still,* move the tube *to remove the wire. Be especially careful when removing the tube not to catch the loop tip on the tube lip. This springing action of the loop creates bacterial aerosols.*

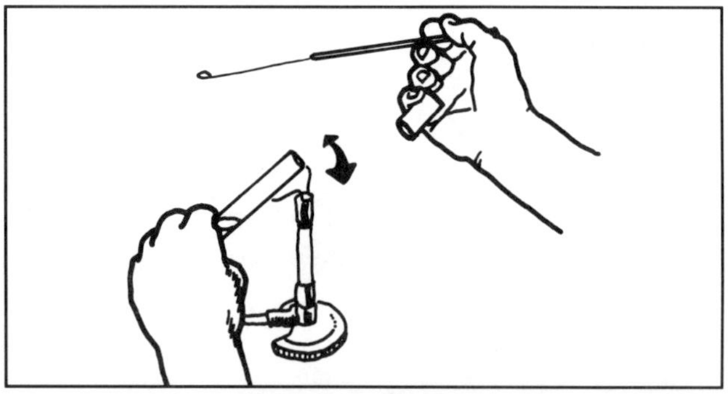

FIGURE B-6g.

Sterilize the Tube Again *Flame the tube lip as before. Keep your loop hand still.*

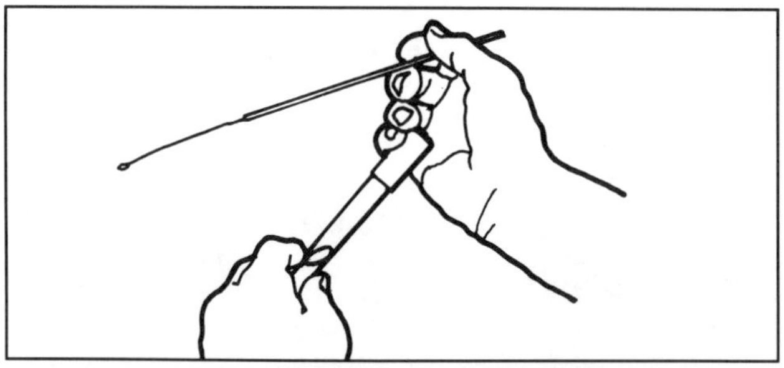

FIGURE B-6h.

Replace the Cap *Keeping the loop hand still (remember, it has growth on it), move the tube to replace its cap. The cap at this point doesn't need to be on firmly — just enough to cover the tube. What you do next depends on the medium to which you are transferring the growth. Please continue with the appropriate inoculation section.*

Transfers From a Broth Culture Using a Pipette

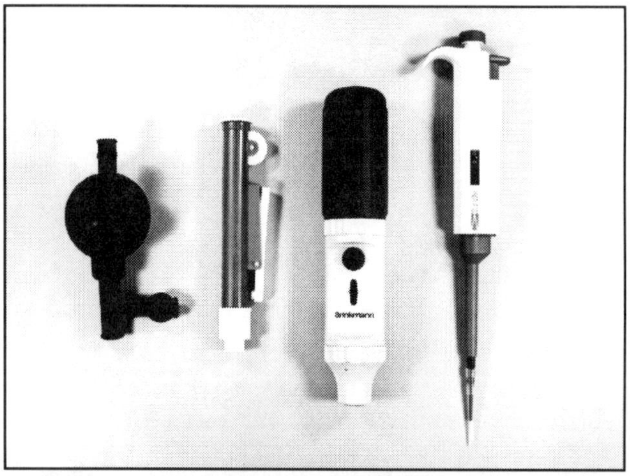

FIGURE B-7.

Mechanical Pipettors *Mouth pipetting is dangerous and has been replaced by mechanical pipettors. Several examples are shown here, each with its own method of operation. Your instructor will show you how to properly use the style of pipettor available in your lab. From left to right: a pipette bulb, a plastic pump, a pipette filler/dispenser, and a micropipettor.*

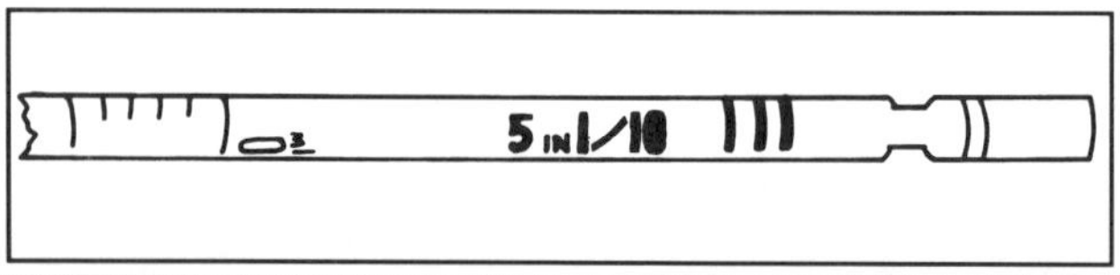

FIGURE B-8a.

Pipette Calibration *Read the pipette calibration. The numbers indicate the pipette's* total volume *and its* smallest calibrated increments. *This is a 5.0 mL pipette divided into 0.1 mL increments.*

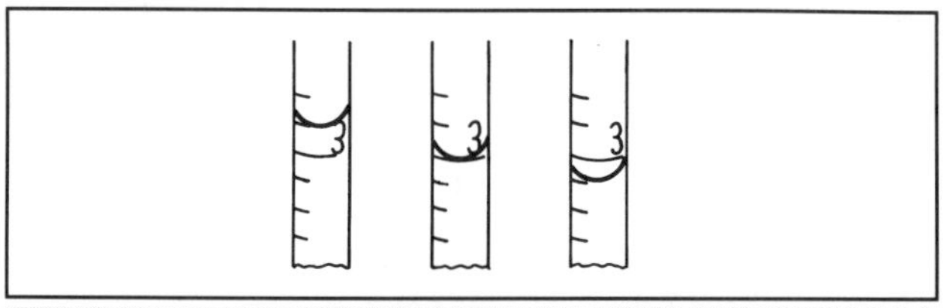

FIGURE B-8b.

Read the Base of the Meniscus *When reading volumes, use the base of the meniscus. The volume in the center pipette is read at exactly 3.0 mL because the meniscus is resting on the line. The left pipette is read at 2.9 mL and the right pipette is read as 3.1 mL. Although the difference in volume between these three pipettes may seem negligible (1 part in 70, a 1.4% error), it may be enough to introduce substantial error into your work.*

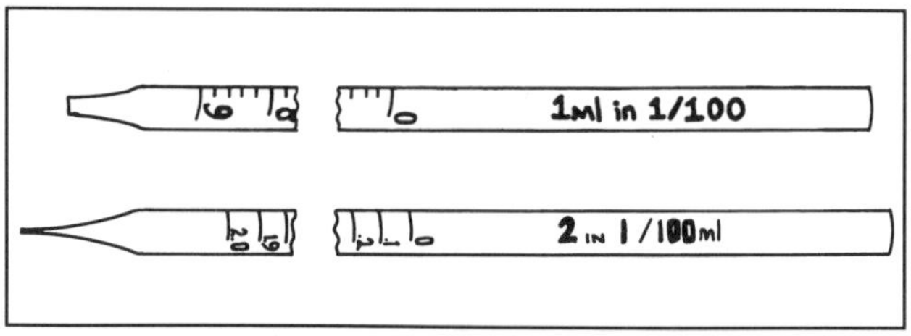

FIGURE B-8c.

Two Types of Pipettes *Two pipette styles are used in microbiology. These are the* serological pipette *(above) and the* Mohr pipette *(below). A serological pipette is calibrated* to deliver *(TD) its volume by completely draining it and blowing out the last drop. The tip of a Mohr pipette is not graduated, so fluid flow must be stopped at a calibration line. Stopping the fluid beyond the last line on a Mohr pipette results in an unknown volume being dispensed. (If pipetting a bacterial culture, be careful not to allow any to drop from the pipette before disposing of it in the autoclave container. Clean up any spills.) In either case, volumes are read at the bottom of the meniscus of fluid.*

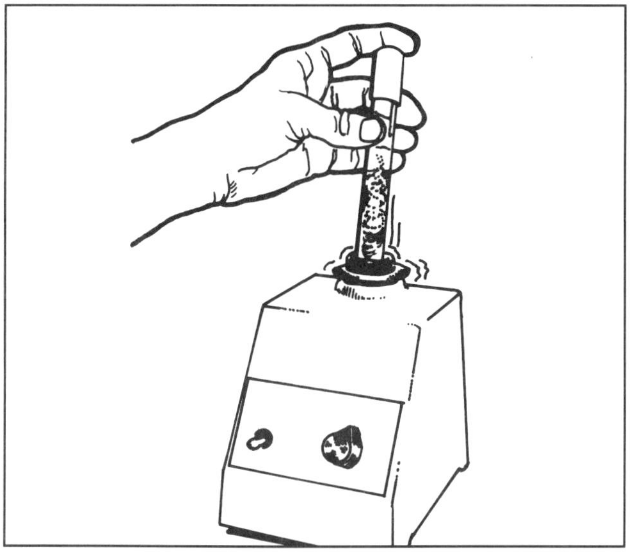

FIGURE B-9a.

Suspend the Bacteria in the Broth (One Method) *Growth may be suspended in the broth with a vortex mixer. Be sure not to mix so vigorously that broth gets into the cap or that you lose control of the tube. It's best to start slowly, then gently increase the speed until the tip of the vortex reaches the bottom of the tube.*

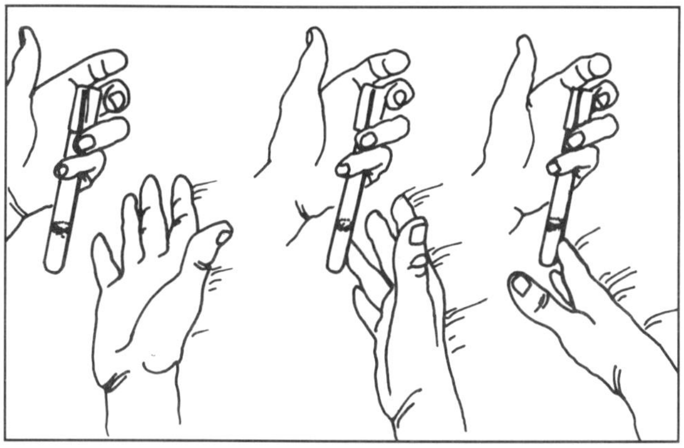

FIGURE B-9b.

Suspend the Bacteria in the Broth (Another Method) *The broth may also be agitated by drumming your fingers along the length of the tube several times. Be careful not to splash the broth into the cap or lose control of the tube.*

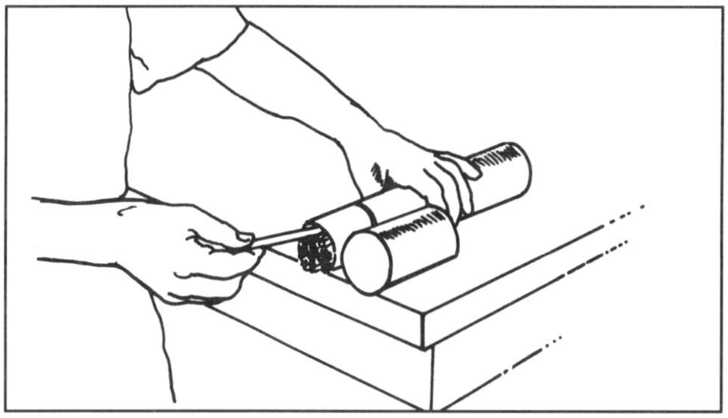

FIGURE B-9c.

Get the Sterile Pipette *Pipettes are sterilized in metal canisters or paper wraps and are stored in groups of a single size.* Be sure you know what volume your pipette will deliver. *Set the canister at the edge of the table and remove its lid. (Canisters should not be stored in an upright position as they may fall over and break the pipettes or become contaminated.) If using pipettes in paper wraps, open the end* opposite the tips. *Grasp* one pipette only *and remove it.*

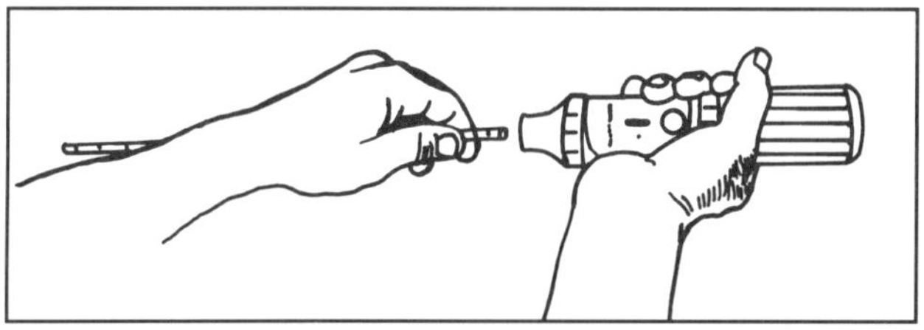

FIGURE B-9d.

Assemble the Pipette *Carefully insert the pipette into the mechanical pipettor. It's best to grasp the pipette near the end with your finger tips. This gives you more control and reduces the chances that you'll break the pipette and cut your hand. Be sure not to touch any part of the pipette that will contact the specimen or the medium or you risk introducing a contaminant. Also, do not lay the pipette on the table top while you continue.*

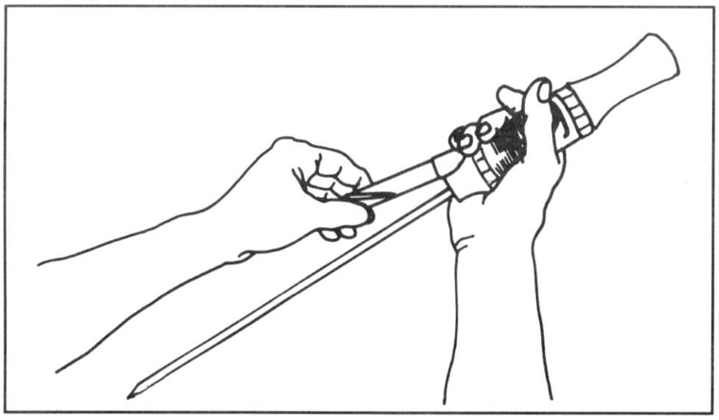

FIGURE B-9e.

Hold the Cap in Your Pinkie Finger *While keeping the pipette hand still, bring the culture tube toward your other hand. Use your pinkie finger to remove and hold its cap. (The cap may be loosened prior to the transfer, especially if it's a screw top cap.)*

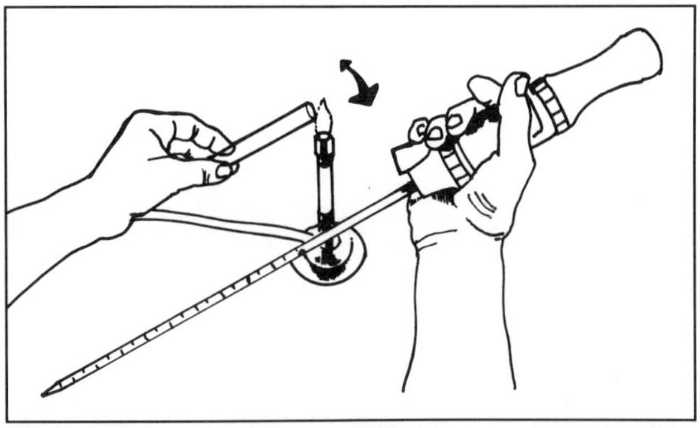

FIGURE B-9f.

Sterilize the Tube *Pass the lip of the tube quickly through the flame two or three times to sterilize the glass and the surrounding air. The tube should be held on a 45° angle to prevent contamination from above. Keep your pipette hand still.*

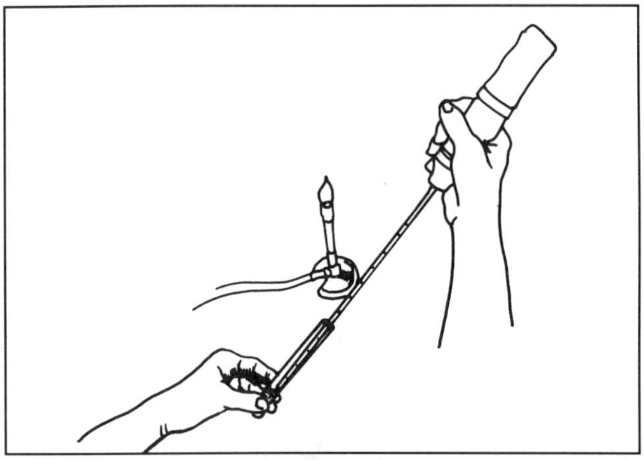

FIGURE B-9g.

Remove the Desired Volume *Insert the pipette and withdraw the appropriate volume. Bring the pipette to a vertical position briefly to accurately read the pipette. (Remember: the volumes in the pipette are correct only if the meniscus of the fluid inside is resting* on *the line, not below it.) Then carefully remove the pipette.*

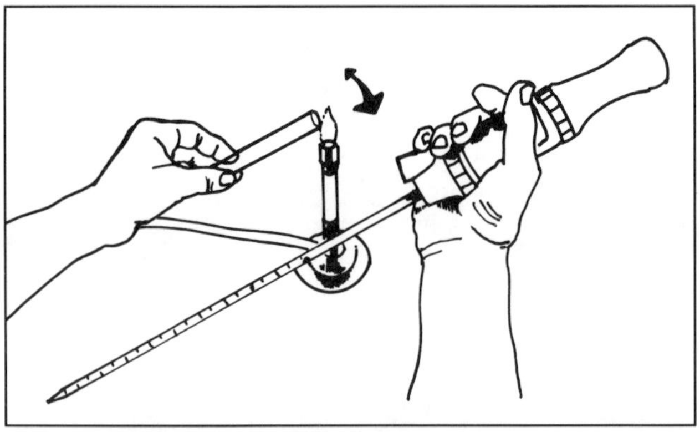

FIGURE B-9h.

Sterilize the Tube Again *Flame the tube lip as before. Keep your pipette hand still.*

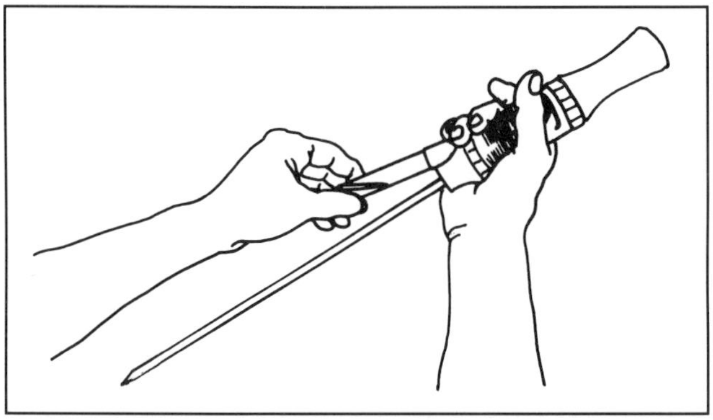

FIGURE B-9i.

Replace the Cap *Keeping the pipette hand still (remember, it contains fluid with microbes in it), move the tube to replace its cap. The cap at this point doesn't need to be on firmly — just enough to cover the tube. What you do next depends on the medium to which you are transferring the growth. Please continue with the appropriate inoculation section.*

Transferring to a Sterile Medium

Inoculation of Agar Plates Using a Cotton Swab

Since the cotton swab was probably used to obtain a specimen containing a mixed culture of microbes, agar plates are the typical medium inoculated. Depending on your purposes, you may inoculate the agar surface in a couple of different ways. A zigzag inoculation is shown in Figure B-10a. Figure B-10b shows inoculation with the swab in preparation for a streak plate.

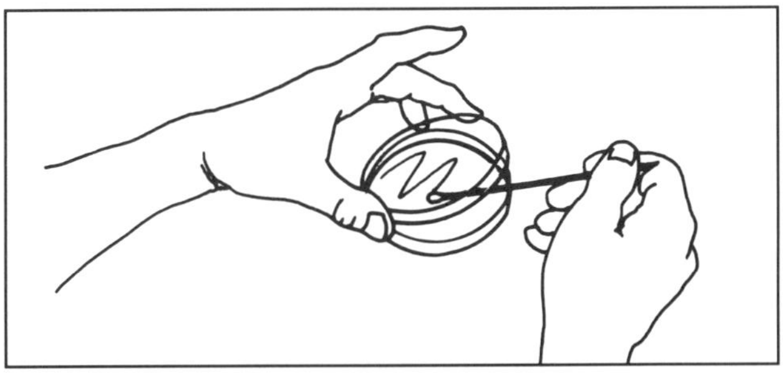

FIGURE B-10a.

Zigzag Inoculation *This inoculation pattern is usually used when the sample does not have a high cell density. Hold the swab comfortably in one hand and lift the lid of the Petri dish with the fingers and thumb of the other. Use the lid as a shield to protect the agar from aerial contamination. Lightly drag the cotton swab across the agar surface in a zigzag pattern. Dispose of the swab as shown in Figure B-10c. Incubate the plate in an inverted position for the assigned time at the appropriate temperature. Be sure to label the base with your name, date and sample.*

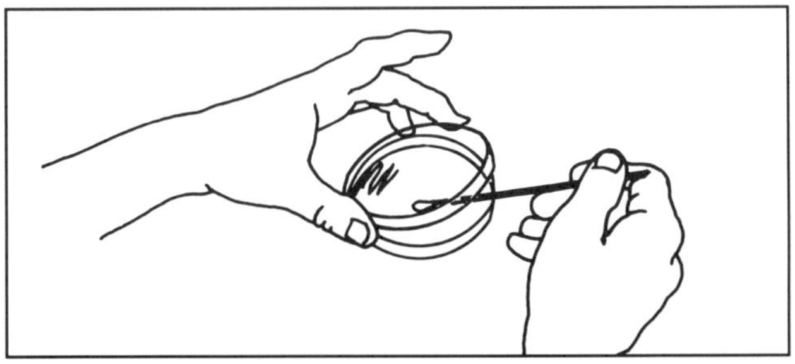

FIGURE B-10b.

Streak Plate Inoculation *This inoculation pattern is usually used when the sample has a high cell density. Hold the swab comfortably in one hand and lift the lid of the Petri dish with the fingers and thumb of the other. Use the lid as a shield to protect the agar from aerial contamination. Lightly drag the cotton swab back and forth across the agar surface in one quadrant of the plate. Further streaking is performed with a loop as described in Figure 2-1, page 8. Dispose of the swab as shown in Figure B-10c. Incubate the plate in an inverted position for the assigned time at the appropriate temperature. Be sure to label the base with your name, date and sample.*

FIGURE B-10c.

Dispose of the Swab *The swab is contaminated and must be disposed of properly in a Biohazard container destined for autoclaving.*

Spot Inoculation of Agar Plates

Sometimes, an agar plate (especially those with differential media) may be used to grow several different specimens at once. Prior to beginning the transfer, the plate may be divided into as many as four sectors using a marking pen. Each may then be inoculated with a different organism. Inoculation involves touching the loop to the agar surface once so that growth is restricted to a single spot — hence the name "spot inoculation."

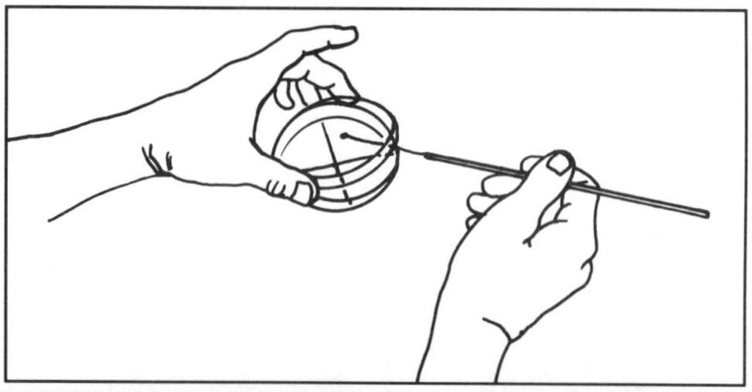

FIGURE B-11a.

Inoculate the Medium *Lift the lid of the sterile agar plate and use it as a shield to prevent aerial contamination Then touch the agar surface in the center of the sector. Remove the loop and replace the lid.*

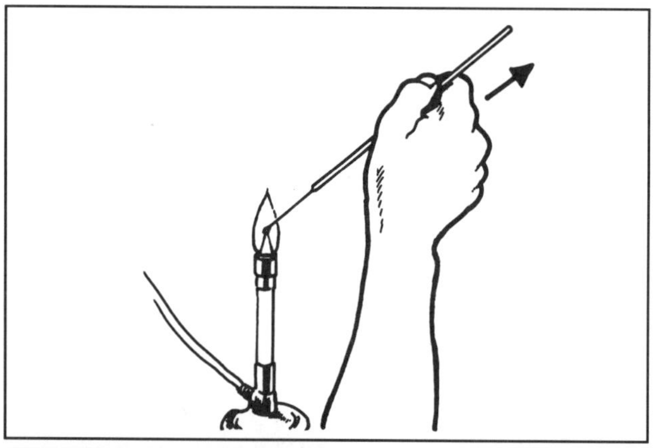

FIGURE B-11b.

Flame the Loop *Sterilize your loop as before. It is especially important to flame it from base to tip now because the loop has lots of bacteria on it. Label the plate's base with your name, date and organism(s) inoculated. Incubate the plate in an inverted position for the assigned time at the appropriate temperature.*

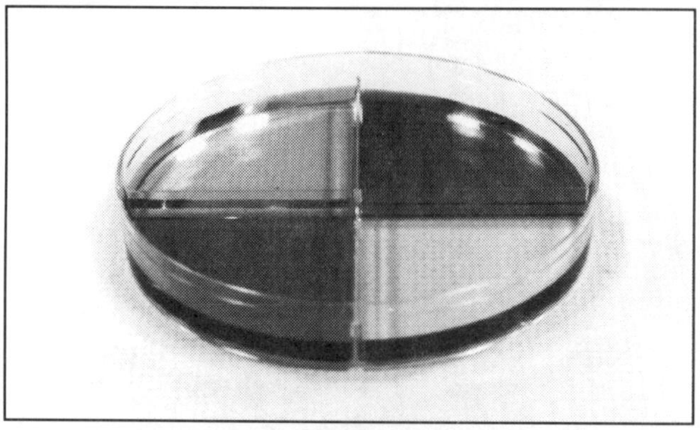

FIGURE B-11c.

Multiple Tests May Be Run *Petri plates may be purchased which have built-in partitions, as in this photo. Four different organisms may be spot inoculated if all wells contain the same medium. Or, as shown here, four different tests can be run simultaneously on the same organism if each well contains a different medium.*

Inoculation of Agar Plates With a Pipette — The Spread Plate Technique

The spread plate technique is often used with quantitative procedures, but may also be used for isolation of a particular organism from a mixed culture.

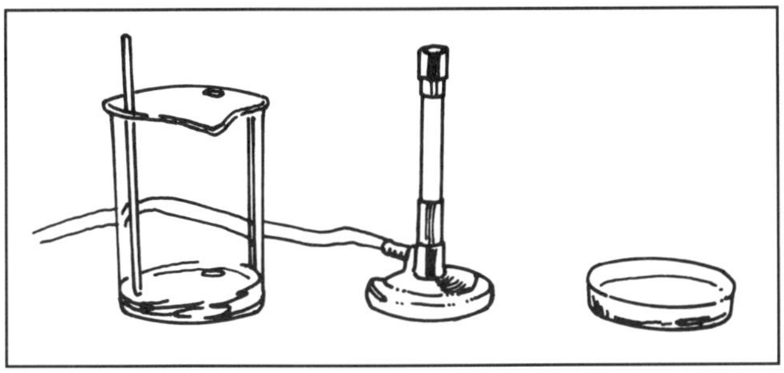

FIGURE B-12a.

Safety First *The spread plate technique requires a Bunsen burner, a beaker with alcohol, a glass spreading rod and the plate. Position these components in your work area as shown: ethyl alcohol, flame, and plate. This arrangement reduces the chance of accidentally catching the alcohol on fire.*

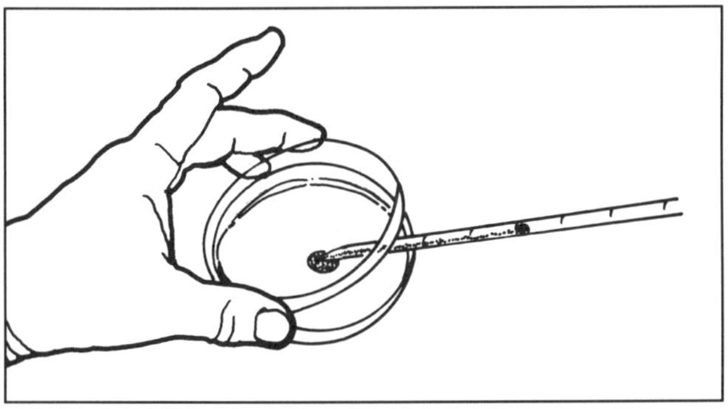

FIGURE B-12b.

Inoculate the Plate *Lift the plate's lid and use it as a shield to protect from airborne contamination. Insert the pipette and dispense the correct volume (often 0.1 mL) onto the center of the agar surface. From this point, the remainder of steps should be completed within about 15 seconds to prevent the inoculum from soaking into the agar.*

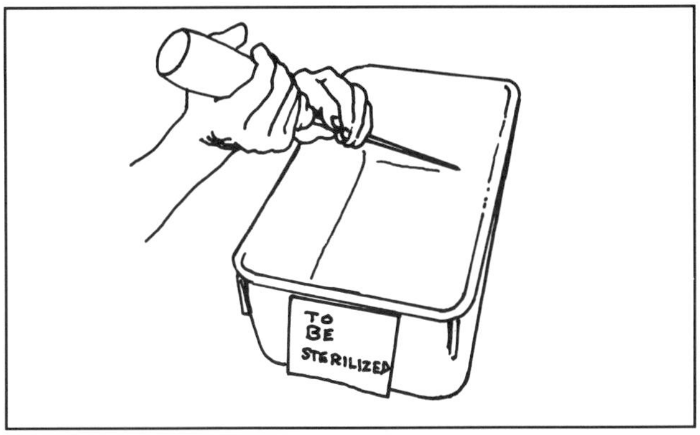

FIGURE B-12c.

Dispose of the Pipette *The pipette is contaminated with microbes and must be correctly disposed of. Each lab has its own specific procedures and your instructor will advise you what to do. Shown here is a glass pipette being placed in a pipette disposal container. When full, the pipettes will be autoclaved. Disposable pipettes must be placed in an appropriate biohazard container. In either case, be careful when removing the pipette from the mechanical pipettor. There is danger of culture dripping from the pipette or of breaking the glass.*

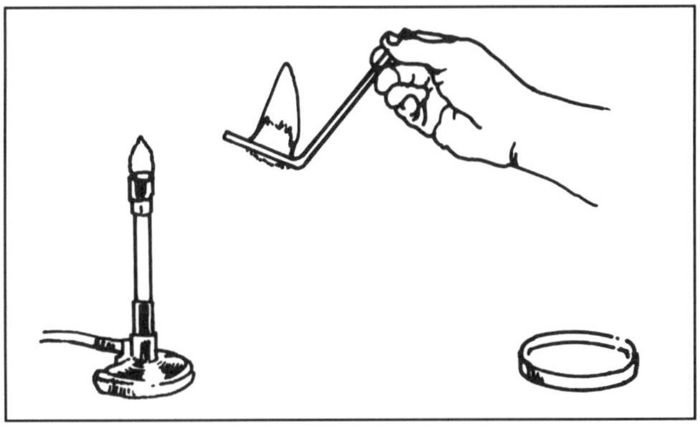

FIGURE B-12d.

Sterilize the Glass Rod *Remove the glass spreading rod from the alcohol and pass it through the flame to sterilize it. Allow the alcohol to burn off the rod.* Do not leave the rod in the flame, *and* be careful not to drop any flaming alcohol on the work surface.

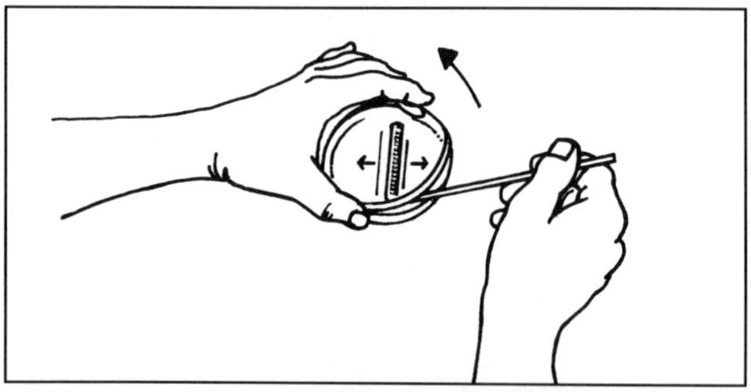

FIGURE B-12e.

Spread the Inoculum *After the flame has gone out on the rod, lift the lid of the plate and use it as a shield from airborne contamination. Then, touch the rod to the agar surface away from the inoculum in order to cool it. To spread the inoculum, hold the plate lid with the base of your thumb and index finger, and use the tip of your thumb and middle finger to rotate the base. At the same time, move the rod in a back-and-forth motion across the agar surface. After a couple of turns, do one last turn with the rod next to the plate's edge.*

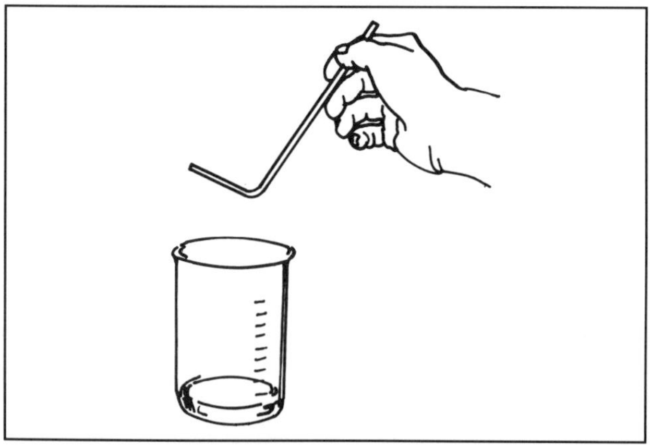

FIGURE B-12f.

Return the Rod to the Alcohol *Remove the rod from the plate and replace the lid. Return the rod to the alcohol in preparation for the next inoculation. Label the plate base with your name, date, organism, and any other relevant information. Incubate the plate in an inverted position at the appropriate temperature for the assigned time. (If you plated a large volume of inoculum, you may need to wait a few minutes to allow it to soak in before inverting the plate.)*

Stab Inoculation of Agar Tubes Using an Inoculating Needle

Stab inoculations of agar tubes are used for several types of differential media. A stab is not used to produce a culture of microbes for transfer to another medium.

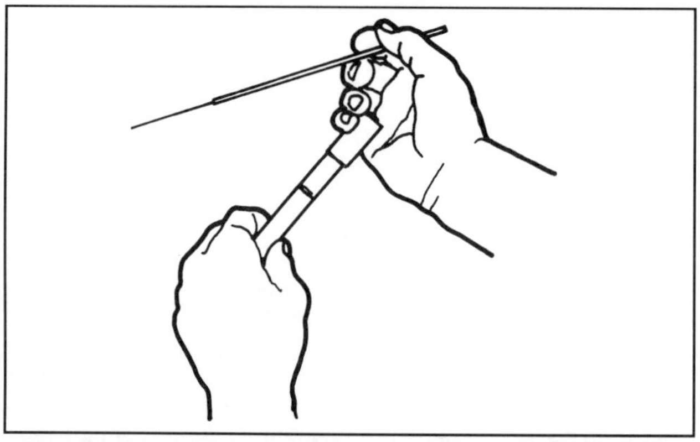

FIGURE B-13a.

Hold the Cap in Your Pinkie Finger *While keeping the needle hand still and away from the flame, bring the agar tube towards your other hand. Use your pinkie finger to remove and hold its cap. (The cap may be loosened prior to the transfer, especially if it's a screw top cap.)*

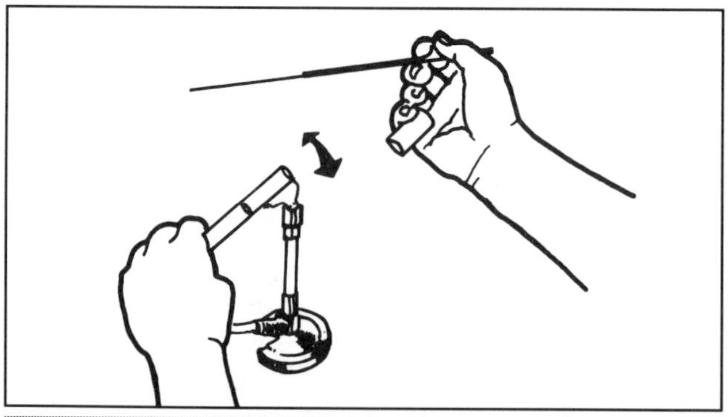

FIGURE B-13b.

Sterilize the Tube *Pass the lip of the tube quickly through the flame two or three times to sterilize the glass and the surrounding air. The tube should be held on a 45° angle to prevent contamination from above. Keep your needle hand still.*

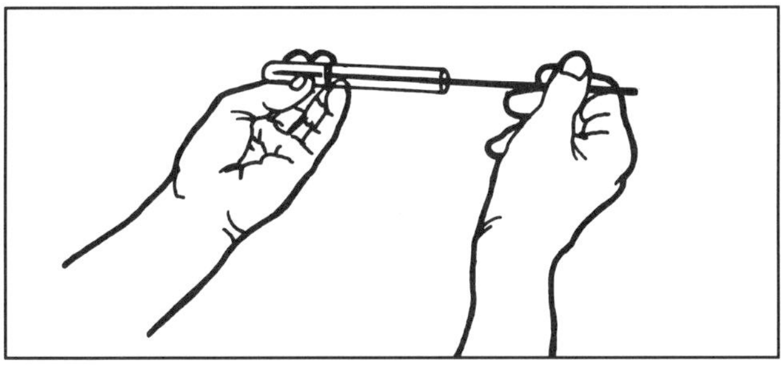

FIGURE B-13c.

Stab the Agar *Hold the open tube at an angle to prevent aerial contamination. Carefully move the agar tube over the needle wire. Insert the needle into the agar, then withdraw the tube carefully (ideally) along the same stab line. Be especially careful when removing the tube not to catch the needle tip on the tube lip. This springing action of the needle creates bacterial aerosols.*

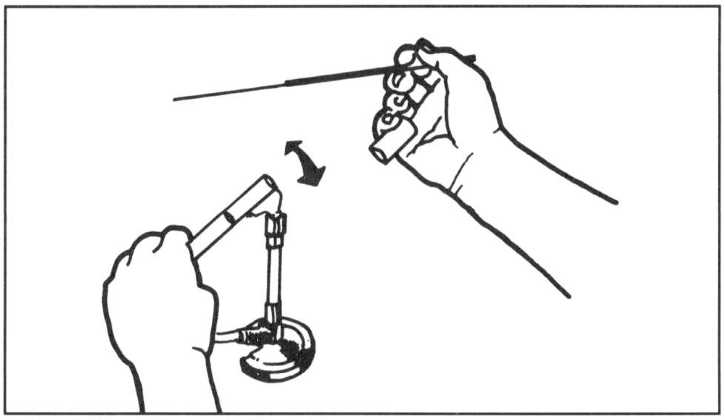

FIGURE B-13d.

Sterilize the Tube Again *Flame the tube lip as before. Keep your needle hand still.*

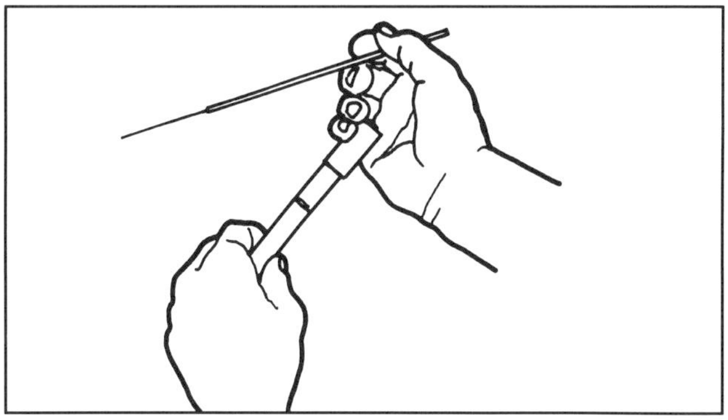

FIGURE B-13e.

Replace the Cap *Keeping the needle hand still (remember, it has growth on it), move the tube to replace its cap. The cap at this point doesn't need to be on firmly — just enough to cover the tube.*

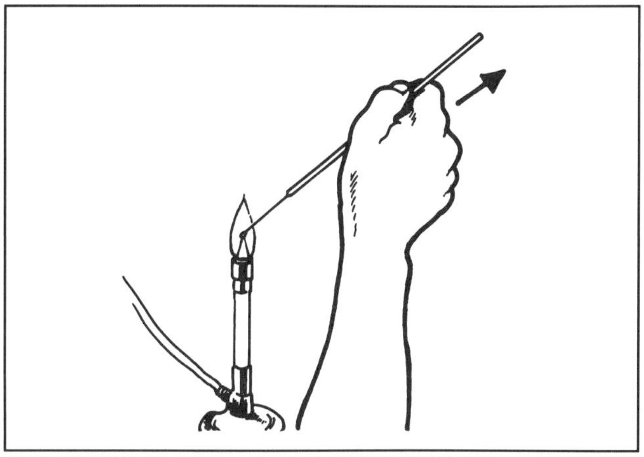

FIGURE B-13f.

Flame the Needle *Sterilize the needle as before by incinerating it in the Bunsen burner flame. Label the tube with your name, date and organism. Incubate at the appropriate temperature for the assigned time.*

Fishtail Inoculation of Agar Slant Tubes

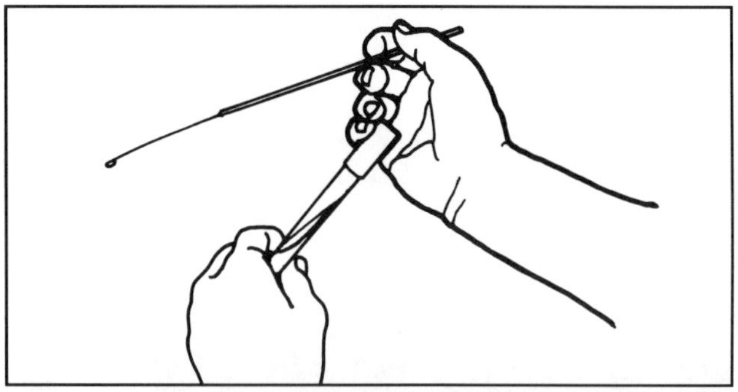

FIGURE B-14a.

Hold the Cap in Your Pinkie Finger *While keeping the loop hand still and away from the flame, bring the agar slant tube toward your other hand. Use your pinkie finger to remove and hold its cap. (The cap may be loosened prior to the transfer, especially if it's a screw top cap.)*

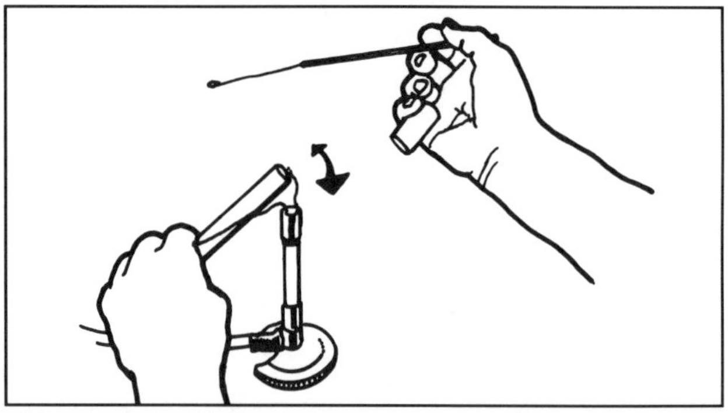

FIGURE B-14b.

Sterilize the Tube *Pass the lip of the tube quickly through the flame two or three times to sterilize the glass and the surrounding air. The tube should be held on a 45° angle to prevent contamination from above. Keep your loop hand still.*

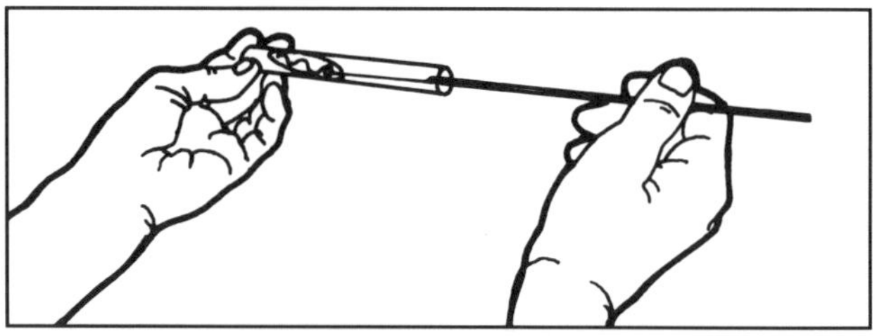

FIGURE B-14c.

Inoculate the Agar *Hold the open tube at an angle to prevent aerial contamination and with the slant surface up. Carefully move the agar tube over the loop. Touch the bottom of the agar surface with the loop. As you remove the tube from over the loop, drag it on the agar surface in a zigzag pattern. Be especially careful when removing the tube not to catch the loop tip on the tube lip. This springing action of the loop creates bacterial aerosols.*

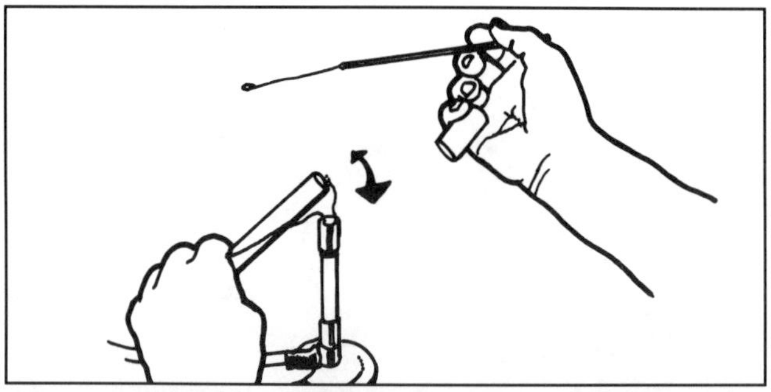

FIGURE B-14d.

Sterilize the Tube Again *Flame the tube lip as before. Keep your loop hand still.*

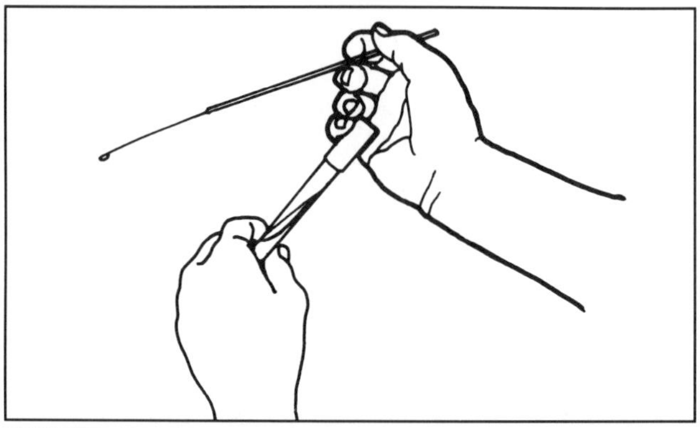

FIGURE B-14e.

Replace the Cap *Keeping the loop hand still (remember, it has growth on it), move the tube to replace its cap. The cap at this point doesn't need to be on firmly —just enough to cover the tube.*

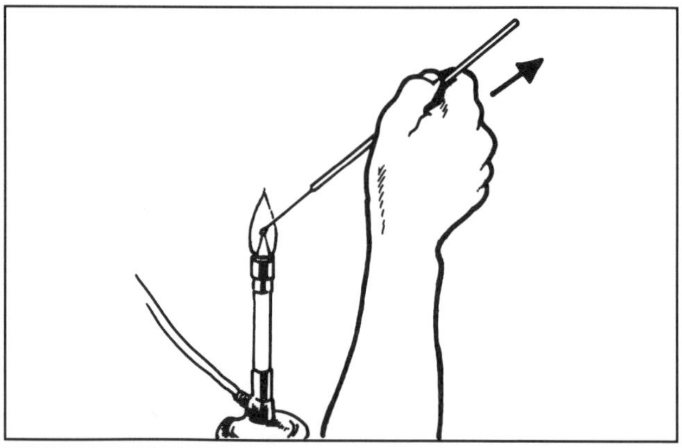

FIGURE B-14f.

Flame the Loop *Sterilize the loop as before by incinerating it in the Bunsen burner flame. Label the tube with your name, date and organism. Incubate at the appropriate temperature for the assigned time.*

Inoculation of Broth Tubes With a Loop or Needle

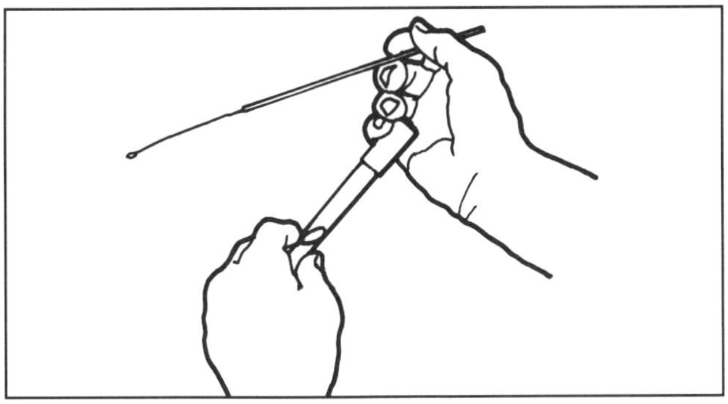

FIGURE B-15a.

Hold the Cap in Your Pinkie Finger *While keeping the loop/needle hand still and away from the flame, bring the broth tube toward your other hand. Use your pinkie finger to remove and hold its cap. (The cap may be loosened prior to the transfer, especially if it's a screw top cap.)*

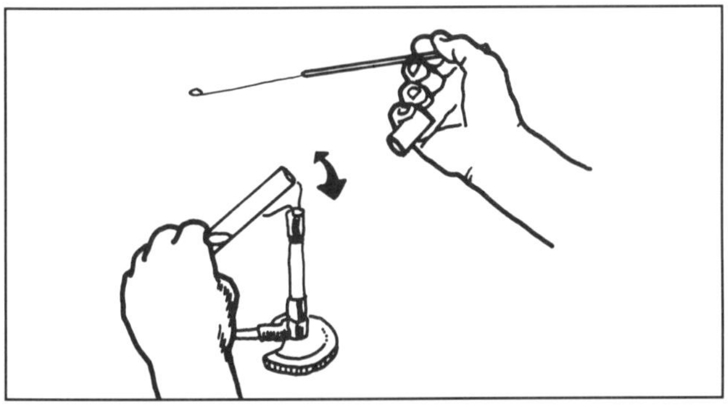

FIGURE B-15b.

Sterilize the Tube *Pass the lip of the tube quickly through the flame two or three times to sterilize the glass and the surrounding air. The tube should be held on a 45° angle to prevent contamination from above. Keep your loop/needle hand still.*

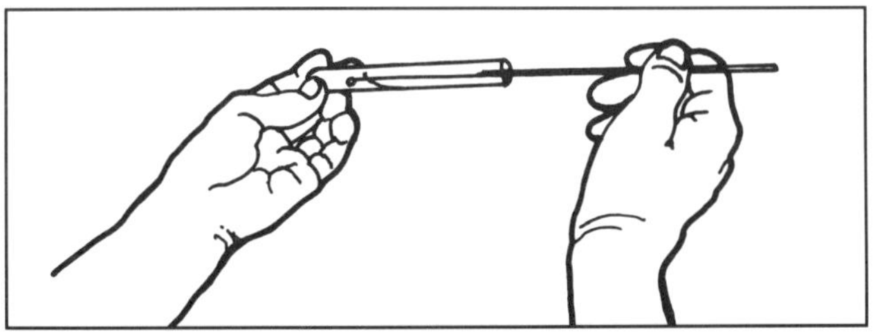

FIGURE B-15c.

Inoculate the Broth *Hold the open tube at an angle to prevent aerial contamination. Carefully move the broth tube over the wire. Gently swirl the loop/needle to dislodge microbes.*

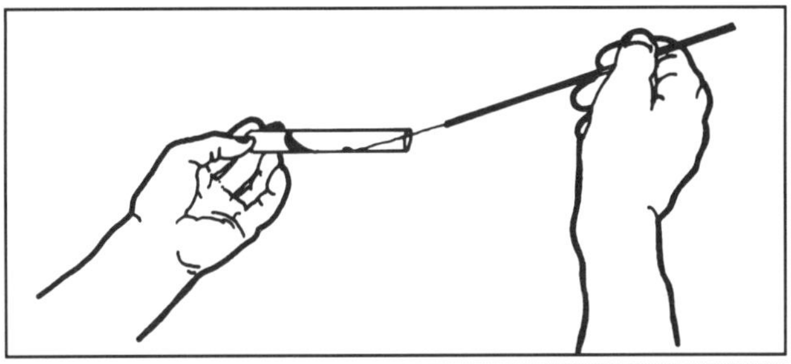

FIGURE B-15d.

Remove Excess Broth *Withdraw the tube from over the loop/needle. Before completely removing it, touch the loop/needle tip to the glass to remove any excess broth. Be especially careful when removing the tube not to catch the loop/needle tip on the tube lip. This springing action of the loop/needle creates bacterial aerosols.*

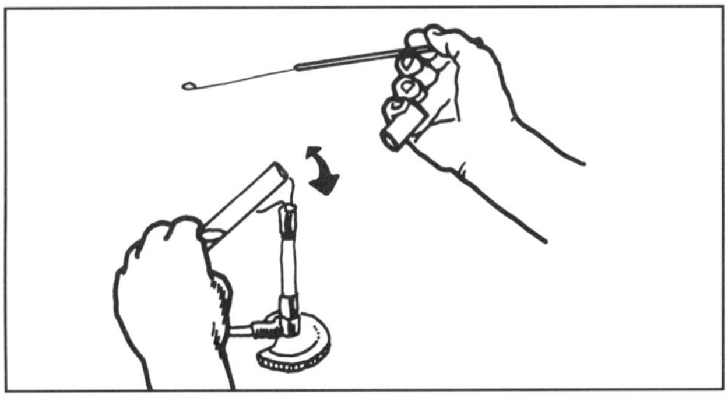

FIGURE B-15e.

Sterilize the Tube Again *Flame the tube lip as before. Keep your loop/needle hand still.*

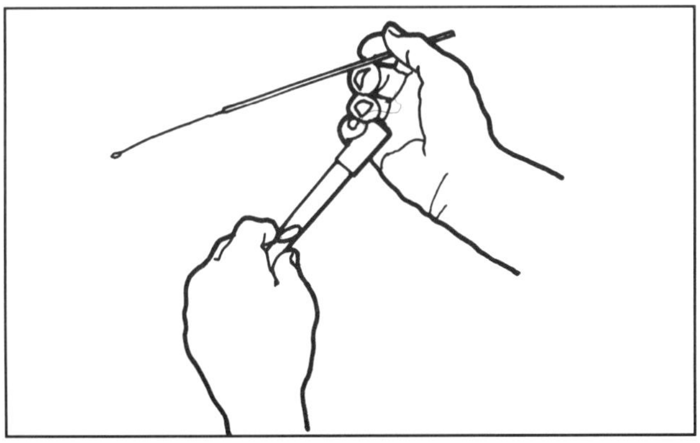

FIGURE B-15f.

Replace the Cap *Keeping the loop/needle hand still (remember, it has growth on it), move the tube to replace its cap. The cap at this point doesn't need to be on firmly — just enough to cover the tube.*

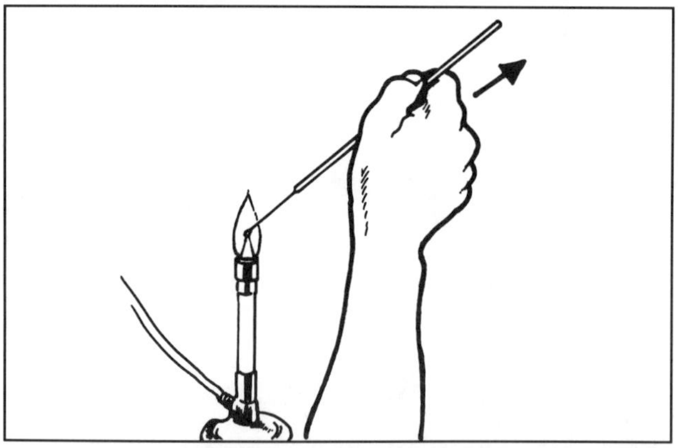

FIGURE B-15g.

Flame the Loop/Needle *Sterilize the loop/needle as before by incinerating it in the Bunsen burner flame. Label the tube with your name, date and organism. Incubate at the appropriate temperature for the assigned time.*

Inoculation of Broth Tubes With a Pipette

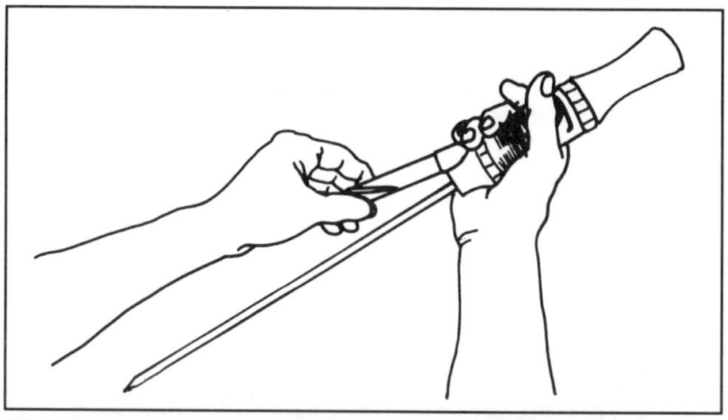

FIGURE B-16a.

Hold the Cap in Your Pinkie Finger *While keeping the pipette hand still and away from the flame, bring the broth tube toward your other hand. Use your pinkie finger to remove and hold its cap. (The cap may be loosened prior to the transfer, especially if it's a screw top cap.)*

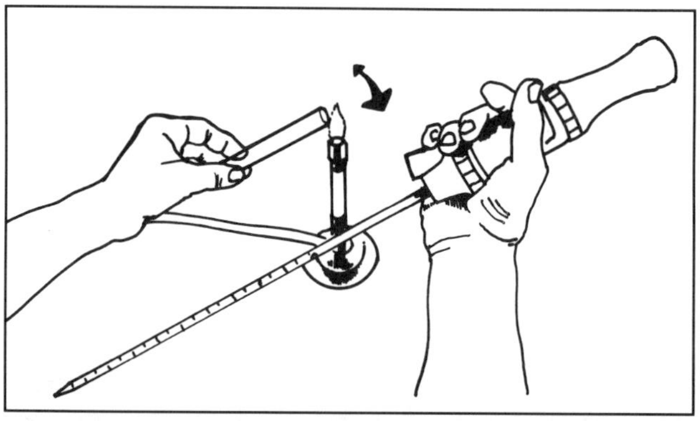

FIGURE B-16b.

Sterilize the Tube *Pass the lip of the tube quickly through the flame two or three times to sterilize the glass and the surrounding air. The tube should be held on a 45° angle to prevent contamination from above. Keep your pipette hand still.*

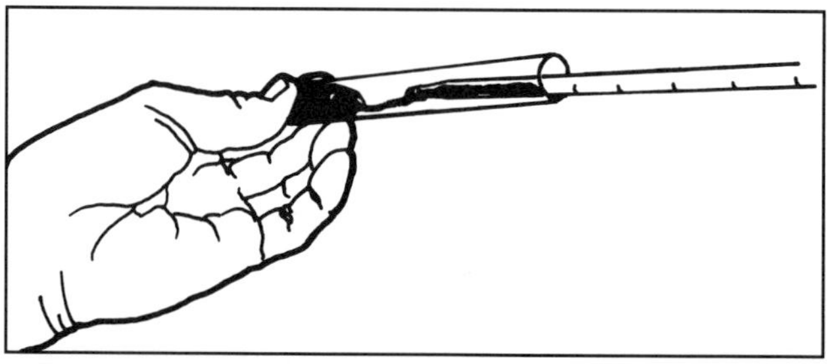

FIGURE B-16c.

Inoculate the Broth *Hold the open tube at a 45° angle to prevent aerial contamination. Insert the pipette tip and dispense the correct volume of inoculum. Blow out the last drop if appropriate.*

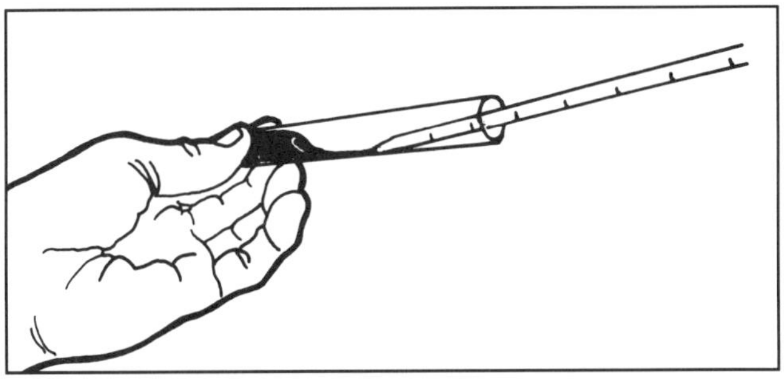

FIGURE B-16d.

Remove Excess Broth *Withdraw the tube from over the pipette. Before completely removing it, touch the pipette tip to the glass to remove the any excess broth. Completely remove the pipette, but avoid waving it around. This can create aerosols.*

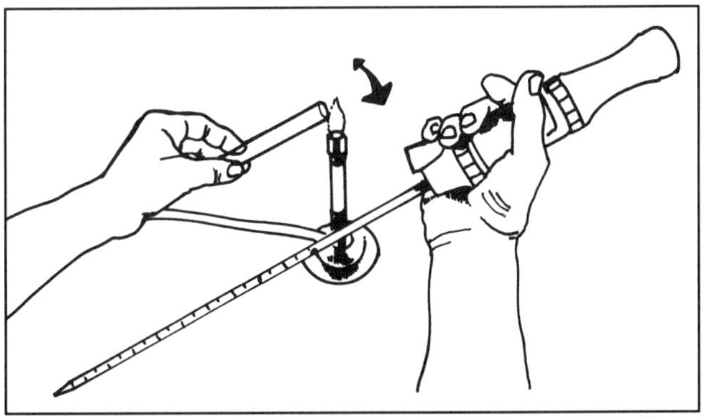

FIGURE B-16e.

Sterilize the Tube Again *Flame the tube lip as before. Keep your pipette hand still.*

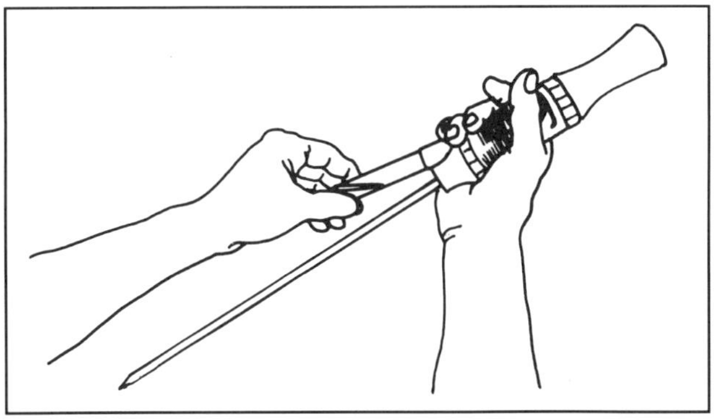

FIGURE B-16f.

Replace the Cap *Keeping the pipette hand still, move the tube to replace its cap. The cap at this point doesn't need to be on firmly — just enough to cover the tube.*

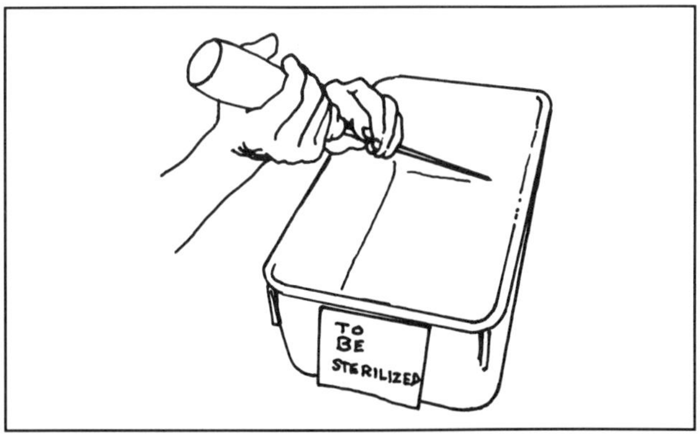

FIGURE B-16g.

Dispose of the Pipette *The pipette is contaminated with microbes and must be correctly disposed of. Each lab has its own specific procedures and your instructor will advise you what to do. Shown here is a glass pipette being placed in a pipette disposal container. When full, the pipettes will be autoclaved. Disposable pipettes must be placed in an appropriate biohazard container. In either case, be careful when removing the pipette from the mechanical pipettor. There is danger of culture dripping from the pipette or of breaking the glass.*

References

Barkley, W. Emmett and John H. Richardson. 1994. Chapter 29 in *Methods for General and Molecular Bacteriology*. American Society for Microbiology, Washington, D.C.

Boyer, Rodney F. 1993. *Modern Experimental Biochemistry, 2nd Ed.* Benjamin/Cummings Publishing Company, Inc., Redwood City, CA.

Claus, G. William. 1989. Chapter 2 in *Understanding Microbes — A Laboratory Textbook for Microbiology*. W.H. Freeman and Company, New York, NY.

Darlow, H. M. 1969. Chapter VI in *Methods in Microbiology, Volume 1*. Edited by J. R. Norris and D. W. Ribbins. Academic Press, Ltd., London.

Fleming, Diane O. 1995. Chapter 13 in *Laboratory Safety — Principles and Practices, 2nd Ed.* Edited by Diane O. Fleming, John H. Richardson, Jerry J. Tulis and Donald Vesley. American Society for Microbiology, Washington, D.C.

Koneman, Elmer W., Stephen D. Allen, William M. Janda, Paul C. Schreckenberger and Washington C. Winn, Jr. Chapter 1 in *Color Atlas and Textbook of Diagnostic Microbiology, 4th Ed.* J.B. Lippincott Company, Philadelphia, PA.

Murray, Patrick R., Ellen Jo Baron, Michael A. Pfaller, Fred C. Tenover, and Robert H. Yolken. 1995. *Manual of Clinical Microbiology, 6th Ed.* American Society for Microbiology, Washington, D.C.

Power, David A. and Peggy J. McCuen. 1988. *Manual of BBL® Products and Laboratory Procedures, 6th Ed.* Becton Dickinson Microbiology Systems, Cockeysville, MD.

APPENDIX C

Light Microscopy

Basic Microscopic Procedures

Proper use of the microscope is absolutely essential for your success in microbiology. Fortunately, with practice and by following a few simple guidelines, you can achieve satisfactory results without much anguish. Since student labs may be supplied with a variety of microscopes, your instructor may supplement the following procedures and guidelines with instructions specific to your equipment. Refer to Figure C-1 as you read the following.

1. Carry your microscope to your work station using both hands — one hand grasping the microscope's arm, the other supporting the microscope beneath its base.
2. Use lens paper to clean the objective, ocular, and condenser lenses. Use a double thickness of paper folded over your index finger to remove the dirt. Do not use force. Even lens paper improperly used can scratch a lens. For stubborn dirt, you may moisten the lens paper with a commercially available lens cleaning solution.
3. Raise the condenser to its maximum position nearly even with the stage and open the iris diaphragm.
4. Plug in the microscope and turn the lamp on. Adjust the light intensity slowly to its maximum.
5. Move the low power objective (usually 4X) into position.
6. Place the slide on the stage in the mechanical slide holder. Center the specimen over the opening in the stage.
7. If using a binocular microscope, adjust the position of the two oculars to match your own interpupillary distance.
8. Use the coarse focus adjustment knob to bring the image into focus. Bring the image into sharpest focus using the fine focus adjustment knob. Then, observe the specimen with your eyes relaxed and slightly above the oculars to allow the images to fuse into one.
9. If you are using a binocular microscope, you may adjust the oculars' focus to compensate for differences in visual acuity of your two eyes.
10. Adjust the iris diaphragm and condenser position to produce optimum illumination, contrast and image.
11. Scan the specimen to locate a promising region to examine in more detail.
12. If you are observing a nonbacterial specimen, progress through the objectives until you see the degree of structural detail necessary for your

purposes. You will need to adjust the fine focus and illumination for each objective. Before advancing to the next objective, be sure to position a desirable portion of the specimen in the center of the field or you risk "losing it" at the higher magnification.

13. If you are working with a bacterial smear, you will need to use the oil immersion lens. Work through the medium (10X), then high dry (40X) objectives, adjusting the fine focus and illumination for each. Before advancing to the next objective, be sure to position a desirable portion of the specimen in the center of the field or you risk "losing it" at the higher magnification.
14. To use the oil immersion lens, rotate the nosepiece to a position midway between the high dry and oil immersion lens. Then, place a drop of immersion oil on the specimen. *Be careful not to get any oil on the microscope or its lenses, and be sure to clean it up if you do.* Rotate the oil lens so its tip is submerged in the oil drop. Focus and adjust illumination to maximize image quality. (Note: Do not move the stage down at this point or you will lose your focal plane. On a properly adjusted microscope the oil lens and the high dry lens have the same focal plane and the oil lens, though longer, should clear the slide without scraping it.
15. When finished, lower the stage (or raise the objective) and remove the slide. Dispose of the slide in a jar of disinfectant.
16. When finished for the day, be sure to do the following:
    a. Center the mechanical stage.
    b. Turn off the light and lower the light intensity to its minimum.
    c. Wrap the electrical cord according to your particular lab rules.
    d. Clean any oil off the lenses, stage, *etc.* Be sure to use only lens paper for cleaning any of the optical surfaces of the microscope.
    e. Return the microscope to its appropriate storage place.

Precautions

⚠ Avoid using too much illumination, or you will lose contrast.

⚠ If you do "lose" your specimen, it is generally faster to go to low power and work your way back to where you were than to search aimlessly at the higher magnification.

⚠ Use caution when focusing with the high dry and oil lenses to avoid driving the lens through the microscope slide. Use only the fine focus knob.

⚠ Oil bubbles or the oil immersion lens tip being above the oil produces a poor quality image. Be certain it is *immersed* in oil.

⚠ If you are unable to get the specimen in focus using the oil immersion lens, you may have the slide in upside down.

References

Chapin, Kimberle. 1995. Chapter 4 in *Manual of Clinical Microbiology, 6th Ed.*, edited by Patrick R. Murray, Ellen Jo Baron, Michael A. Pfaller, Fred C. Tenover, and Robert H. Yolken. American Society for Microbiology, Washington, D.C.

Murray, R.G.E and Carl F. Robinow. 1994. Chapter 1 in *Methods for General and Molecular Bacteriology*, edited by Philipp Gerhardt, R.G.E. Murray, Willis A. Wood, and Noel R. Krieg. American Society for Microbiology, Washington, D.C.

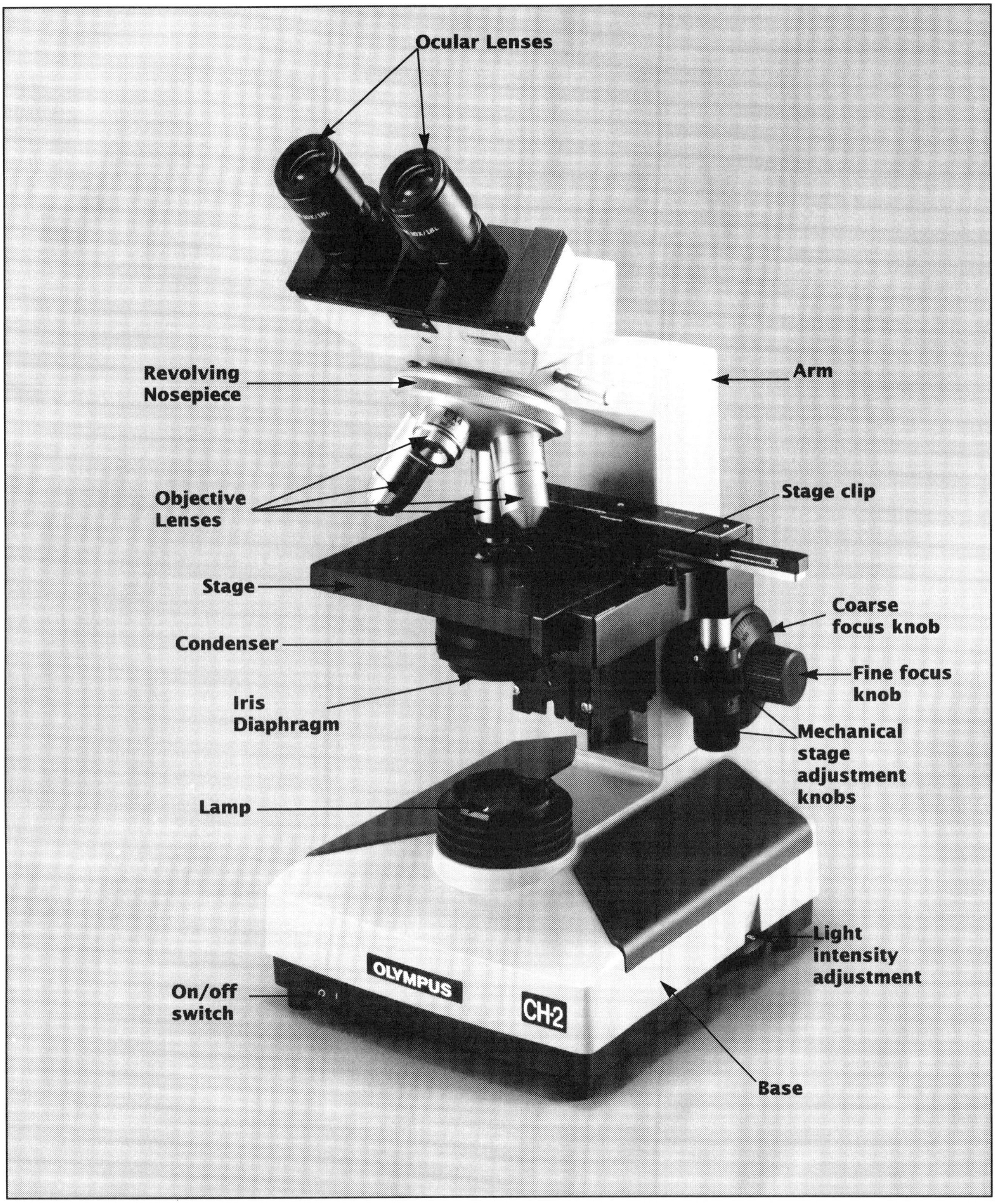

FIGURE C-1.

A Binocular Compound Microscope *A quality microscope is essential to the field of microbiology. (Photograph courtesy of Olympus America, Inc.)*